*Southern Lesser Antilles Arc Platform:
Pre-Late Miocene Stratigraphy, Structure,
and Tectonic Evolution*

R. C. Speed
Department of Geological Sciences
Northwestern University
Evanston, Illinois 60208

P. L. Smith-Horowitz
Division of Geomechanics
Box 54
Mount Waverly
Victoria 3149, Australia

K.v.S. Perch-Nielsen
Geologisches Institut
ETH
Zurich, CH8092, Switzerland

J. B. Saunders
Naturhistorisches Museum
Basel, CH4001, Switzerland

A. B. Sanfilippo
Scripps Institution of Oceanography
University of California
La Jolla, California 92093

SPECIAL PAPER
277
1993

Copyright © 1993, The Geological Society of America, Inc. (GSA). All rights reserved. GSA grants permission to individual scientists to make unlimited photocopies of one or more items from this volume for noncommercial purposes advancing science or education, including classroom use. Permission is granted to individuals to make photocopies of any item in this volume for other noncommercial, nonprofit purposes provided that the appropriate fee ($0.25 per page) is paid directly to the Copyright Clearance Center, 27 Congress Street, Salem, Massachusetts 01970, phone (508) 744-3350 (include title and ISBN when paying). Written permission is required from GSA for all other forms of capture or reproduction of any item in the volume including, but not limited to, all types of electronic or digital scanning or other digital or manual transformation of articles or any portion thereof, such as abstracts, into computer-readable and/or transmittable form for personal or corporate use, either noncommercial or commercial, for-profit or otherwise. Send permission requests to GSA Copyrights.

Copyright is not claimed on any material prepared wholly by government employees within the scope of their employment.

Published by The Geological Society of America, Inc.
3300 Penrose Place, P.O. Box 9140, Boulder, Colorado 80301

Printed in U.S.A.

GSA Books Science Editor Richard A. Hoppin

Library of Congress Cataloging-in-Publication Data

Southern Lesser Antilles arc platform : pre-Late Miocene stratigraphy,
   structure, and tectonic evolution / R.C. Speed . . . [et al.].
     p. cm. — (Special paper ; 277)
   Includes bibliographical references.
   ISBN 0-8137-2277-2
   1. Geology—Antilles, Lesser. I. Speed, Robert C. II. Series:
Special papers (Geological Society of America) ; 277.
QE226.A53S68   1993
551.7'87'09729—dc20   93-897
                                                                                                                  CIP

Front cover: Panorama north-northeastward from Carriacou across southern Lesser Antilles arc platform. Union Island in left foreground. Prune (Palm) Island in right foreground (shoals surrounding two dark hills). Mayreau in center background. Canouan in right background with 300-m-high peak. **Back cover:** West flank of Baradel island in the Tobago Cays, featuring Baradel thrust zone. Hangingwall on left (light hues, Baradel chert unit) above northwest-dipping fault zone and footwall (dark volcanigenic sedimentary rock unit); width of view about 100 m.

10   9   8   7   6   5   4   3   2   1

Contents

Abstract 1

Introduction 2

Regional Structure and Tectonics 2
    Plates and Seismicity 2
    Southern Lesser Antilles Arc System 5

Structure of the Arc Platform 8
    Morphology 8
    Crustal Structure 8
    Flank Structures 8
    Crestal Structure 9
    Neogene Magmatism 10

Geology of the Islands 16
    Introduction 16
    Grenada 17
    Carriacou 26
    Union Island 59
    Prune Island 61
    Jamesby 63
    Baradel 63
    Mayreau 66
    Canouan 72
    Mustique 77

Synthesis of Island Geology 79
    Stratigraphy 79
    Magmatism 85
    Sedimentary Depositional Environments 87
    Deformation 89

Tectonics 92
    Six-Stage Evolution of the SLAAP 92
    Regional Correlations and Interpretations 93
    Paleogene Magmatic Arc 95
    Uplift, Westerly Contraction, Neogene Magmatism 96

Acknowledgments 97

References Cited 97

Geological Society of America
Special Paper 277
1993

Southern Lesser Antilles Arc Platform: Pre-Late Miocene Stratigraphy, Structure, and Tectonic Evolution

ABSTRACT

We present a study of older rocks exposed on Grenada and the Grenadine Islands of the southern Lesser Antilles arc platform (SLAAP) with a view toward understanding the structural and stratigraphic evolution of the platform that led to the modern (and Neogene) magmatic arc. The modern SLAAP is an east-tilted half horst that probably developed early in Miocene time by uplift along a zone of north-striking normal faults relative to both sea level and the crust of the adjacent Grenada Basin. The uplift was completed before Neogene magmas erupted at the surface of the SLAAP beginning at about 12 Ma.

Twenty-six units of older rocks, defined as those that preceded local Neogene magmatism, have been identified by us and previous workers in the SLAAP. Our contributions are new stratigraphic and structural analyses and microfossil and radiometric dating. The age range of older rocks is early middle Eocene to middle Miocene. The most stratigraphic information comes from Carriacou, which contains newly defined formations of Eocene and Oligocene age that are thrust above markedly different late Eocene facies. The thrust is covered by Miocene, possibly late Oligocene, strata below which there is a lacuna possibly as long as 10 m.y. The oldest rocks of the SLAAP are the middle Eocene Mayreau Basalt. Earlier claims of a Cretaceous age of rocks on Union Island are in error; we show them to be Eocene. Assuming they were all deposited contiguously, the older rocks units can be grouped and interpreted as follows: I, middle Eocene pillow basalt of spreading origin and pelagic cover; II, late middle Eocene–middle Miocene deep basinal sediments comprising arc-volcanigenic turbidite and hemipelagite, together with minor intrusive basalt; III, early(?) and middle Miocene local carbonate platform. Group I basalts are correlated with crust of the Grenada Basin, and the Eocene and Oligocene sediments of I and II are equivalent to the deep strata of the Grenada Basin. No magmatic arc rocks of Paleogene age are recognized in the SLAAP.

Older rocks of the SLAAP are deformed by folding, thrusting, and foliation development, and by Neogene normal faulting and intrusion. The principal deformation was a horizontal contraction of northerly bearing in late Oligocene and (or) early Miocene time before the uplift of the half horst. A later horizontal contraction of westerly bearing occurred in the middle Miocene after the horst had developed. The northerly contraction in the SLAAP is interpreted to have arisen by accretion of Grenada Basin cover and slices of shallow basement during a brief episode of subduction of the oceanic crust of the Grenada Basin relatively southward below the (?) Tobago terrane. The accretionary prism of the SLAAP is inferred to have been continuous with the accretionary belt of the Southern Grenada Basin Deformation Front that extends west-southwest to a point west of Margarita.

A Paleogene magmatic arc was the source of copious volcanigenic and carbonate clastic sediment to the SLAAP basin in late Middle Eocene and Oligocene times. This

Speed, R. C., Smith-Horowitz, P. L., Perch-Nielsen, K.v.S., Saunders, J. B., and Sanfilippo, A. B., 1993, Southern Lesser Antilles Arc Platform: Pre-Late Miocene Stratigraphy, Structure, and Tectonic Evolution: Boulder, Colorado, Geological Society of America Special Paper 277.

arc was not in the SLAAP and has not been directly located. We infer its locus is south of the Grenada Basin, striking west-southwest from Grenada; its original orientation is uncertain, but scanty paleomagnetic data provide no evidence for large rotation.

The Neogene magmatic arc of the southern Lesser Antilles is not a clone of the Paleogene arc. It developed transverse to the Paleogene arc and SLAAP accretionary prism, presumably by a major reconfiguration within the Caribbean-American plate boundary zone. The large change in arc trend is thought to be due to collision between the Paleogene arc system and continental South America.

INTRODUCTION

We present new field studies and interpretations of older (pre-late Miocene) rocks of the southern Lesser Antilles arc platform, exposed on Grenada and some of the Grenadine islands from Mustique south (Figs. 1A, B).

The Lesser Antilles arc platform is a ridge defined bathymetrically between deep marine basins (forearc and backarc) from Grenada north to the Greater Antilles. The western margin of this reach of the arc platform is surmounted by a chain of active and Neogene (including Holocene) volcanoes, for which the Lesser Antilles is best known (Fig. 1). The arc platform, moreover, continues southwest of Grenada as far as Margarita Island between the marine Grenada Basin and sediment-filled Carupano Basin, as indicated by large positive free-air gravity anomalies (Fig. 2). Our geographic focus is a part of the southern Lesser Antilles arc platform (SLAAP) between about 12 and 13° N, including Grenada and some of the Grenadine islands (Fig. 1).

Older rocks of the SLAAP, mainly sedimentary and of Eocene to Miocene age, serve as the foundation to the Neogene volcanic chain. The objectives of this paper are an improved understanding from the older rocks of the structural and stratigraphic evolution of the SLAAP that led up to Neogene arc volcanism. Correlative goals are the whereabouts of a Paleogene magmatic arc and the Paleogene and Miocene tectonics of the southeastern Caribbean.

Our work includes studies in the field of rocks, stratigraphy, and structure (Speed and Smith); micropaleontologic (Perch-Nielson, Sanfilippo, and Saunders) and radiometric dating; and seismic section analyses (Smith). New results, synthesized with earlier contributions by Martin-Kaye (1958, 1969), Robinson and Jung (1972), and Westercamp and other (1985), are here presented in four parts: (1) regional geology and crustal structure; (2) geology of islands (Fig. 1B: Grenada, Carriacou, Union, Prune, Mayreau, Jamesby, Baradel, Canouan and Mustique); (3) synthesis of the depositional, magmatic, bathymetric, and deformational history of the older rocks; and (4) tectonic interpretations.

By way of introduction, our major findings are the following. Older rocks of SLAAP can be divided into three groups, assuming the rocks, now mainly in fault-bounded bodies, evolved as a stratigraphic succession. The lowest group consists of Eocene basalt of spreading origin and Eocene pelagic cover. An intermediate group contains Eocene to Miocene basinal arc-derived sediment gravity flows and hemipelagic strata but no clearly arc-magmatic rocks. The highest group, only of local extent, is constituted by Miocene platformal strata. The older rocks underwent late Oligocene and (or) early Miocene contraction, including thrusting of probable large displacement. They partly emerged from a basinal setting in the Paleogene to a platformal one by middle Miocene time. A Paleogene magmatic arc existed near the older rocks basin but not in what is now the SLAAP (Fig. 1B). The emergence of the SLAAP in the Miocene was earlier than the beginnings of volumetrically major extrusion in today's magmatic arc. These findings are counter to assumptions commonly employed in the tectonic histories and restorations of the Lesser Antilles arc (Martin-Kaye, 1969; Tomblin, 1975; Pindell and others, 1988; Robertson and Burke, 1989).

REGIONAL STRUCTURE AND TECTONICS

Plates and seismicity

The southeastern Caribbean region contains two, possibly three, major plates: the Caribbean Plate (Ca), whose nucleus in this region is the oceanic Venezuelan Basin lithosphere (Fig. 2a), and the South American (SA) and North American (NA) Plates. The position and kinematics of the boundary between SA and NA are uncertain (Speed and others, 1991; Argus, 1990). Because of this and the small relative velocity between SA and NA in the southeastern Caribbean now and through most of the Cenozoic (Ladd, 1976; Klitgord and Schouten, 1986; Pindell and others, 1988; DeMets and others, 1990), we refer hereafter to the American Plate (Am) rather than to SA or NA.

By all modern accounts, the Caribbean Plate moves with an easterly component relative to Am, although the sense and amount of obliquity and the velocity magnitude are disputed (Molnar and Sykes, 1969; Jordan, 1975; Sykes and others, 1982; DeMets and others, 1990). From St. Lucia north (Fig. 1a), seismicity defines well a subduction zone between the downgoing slab of Am and the overriding Ca (Stein and others, 1982; Wadge and Shepherd, 1984). There, the Lesser Antilles arc surmounts the leading edge of Ca and probably moves with the Venezuelan Basin apart from some extensional deformation in intervening tracts (Speed and Walker, 1991).

From St. Lucia south, however, the plate kinematics are relatively obscure, owing to patchy and generally low seismicity and to complicated structure (Stein and others, 1982; Perez and

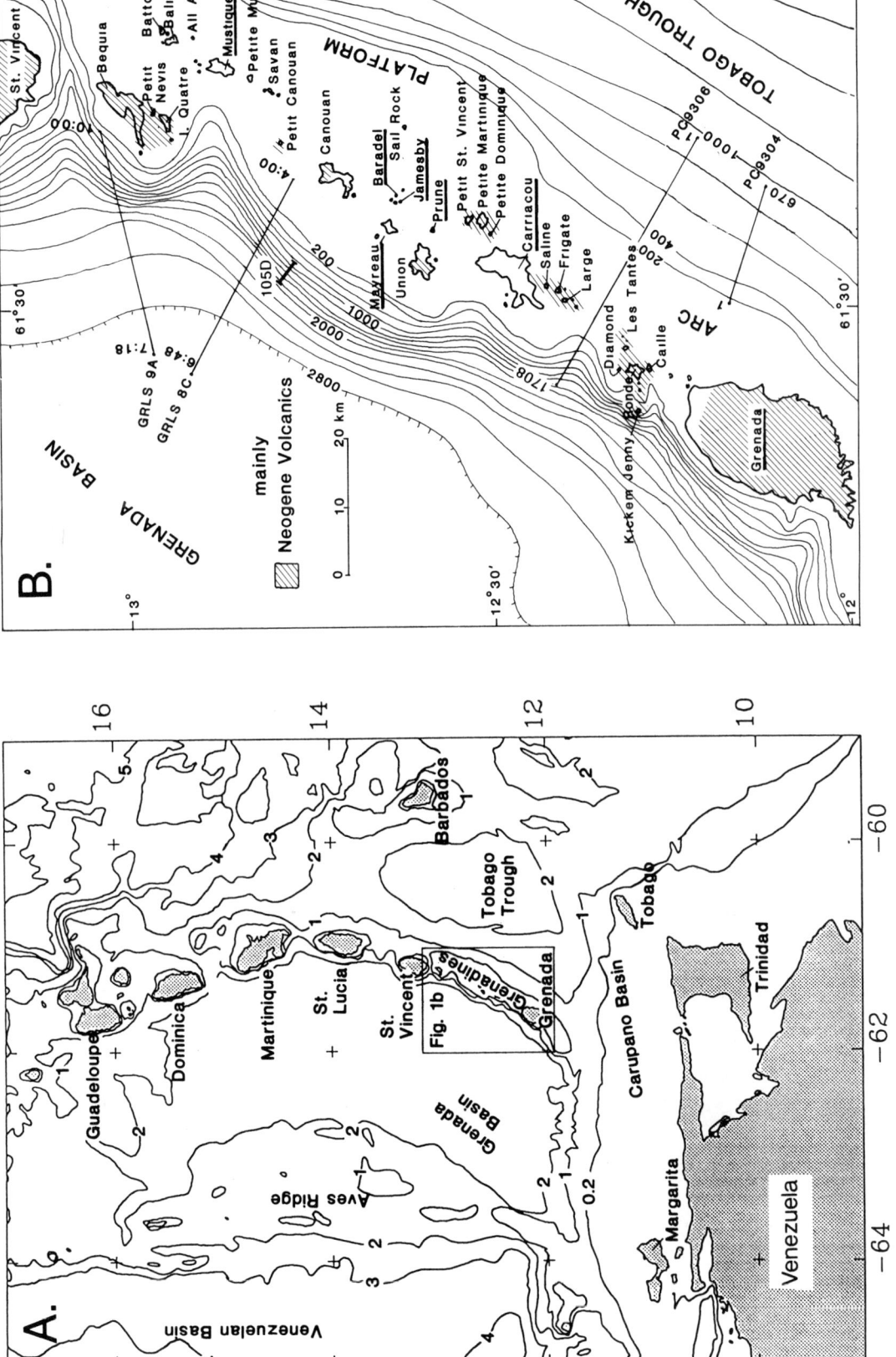

Figure 1. A) Geography and bathymetry of the southeastern Caribbean; land areas patterned; contours in kilometers; B) southern Lesser Antilles arc platform and adjacent basins; islands with underlined names were studied in this research; bathymetric contours in meters from Speed and others (1984); tracks are shown of four seismic sections displayed in Figure 4; 105 D is a dredge haul site; lined areas are mainly Neogene volcanic rocks.

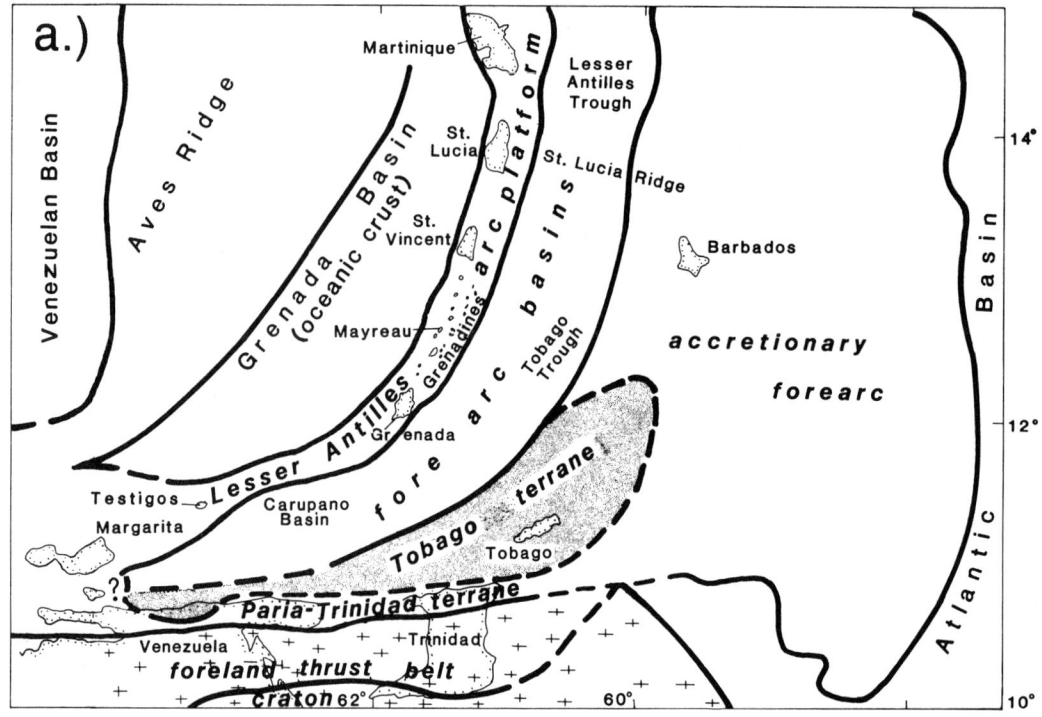

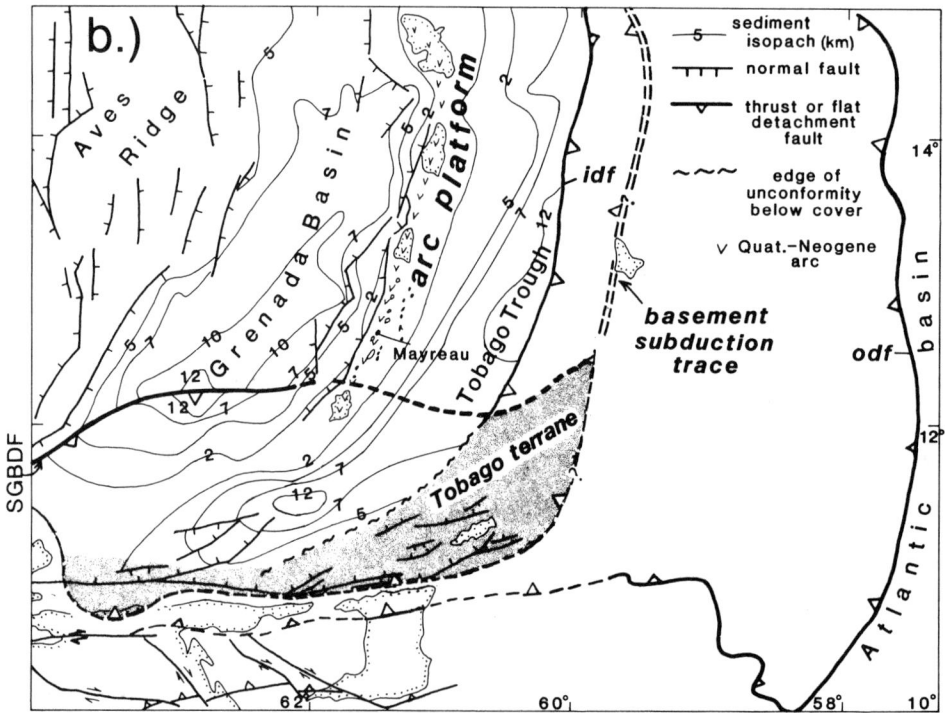

Figure 2 (on this and facing page). Tectonic features and gravity of the southeastern Caribbean. a.) Major tectonic divisions of the southern Lesser Antilles arc and South American continent; dashed lines indicate poorly located, mainly buried boundaries. b.) Major structural features; double-lined thrust is probable trace of subduction between crusts below sedimentary forearc; thrust traces with thin lines are within supracrustal rocks; contours show thickness of sediments in kilometers; idf is inner deformation front and odf is outer deformation front of accretionary forearc. c.) Free air gravity; contours 50-mgal spacing, from Speed and others (1984).

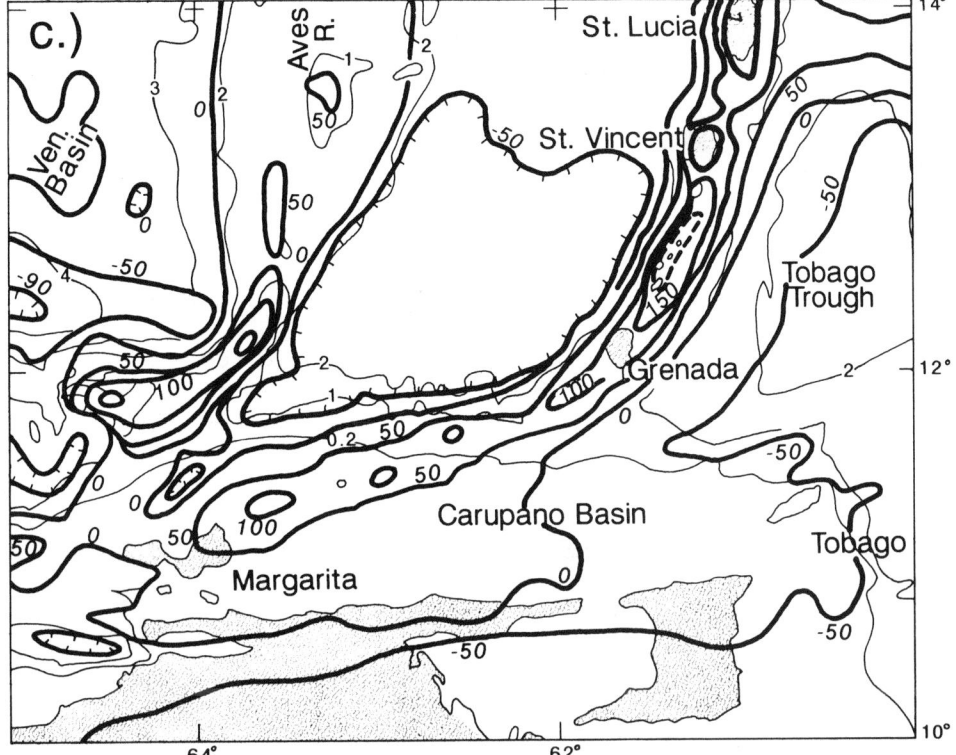

Aggarwal, 1981; Shepherd and others, 1990; Russo and Speed, 1992; Russo and others, 1992). The SLAAP is in an earthquake gap, marked by infrequent scattered seisms at depths between 100 and 160 km with varied focal mechanisms (Russo and others, 1992).

Between the latitudes of St. Lucia and Barbados, the trace of the basement subduction zone can be located by the Bouguer gravity minimum and the coincident structural high of the accretionary forearc (Westbrook, 1975), assuming such features mark the axis of a filled trench. South and west of Barbados, the subduction trace is more difficult to find; it lies in a latitude band between 10.0 and 12.2° N, as indicated by the paucity of deformation in shallow strata north and south of the band. A steeply northwest-dipping slab can be defined in the eastern Carupano Basin (Fig. 2a) by seismicity below about 50 km depth (Russo and Speed, 1992). The displacement field at shallower levels within the band, however, is uncertain; there could be one principal but aseismic displacement surface, or alternatively, displacement could be distributed with a continuous gradient across the band. Some candidates for major fault zone in the band are the Southern Grenada Basin Deformation Front (SGBDF; Fig. 2b), North Coast fault zone along the northern edge of South America, and the foreland thrust belt (Fig. 2a). Perhaps, the best possibility for a major displacement surface is the southern boundary of the Tobago terrane (Fig. 2b) because it joins tracts of greatly dissimilar rocks.

The most probable relative velocity of Ca-Am in the southeastern Caribbean is east-southeast, 1.3 cm/yr. This is calculated from an Euler pole for Ca-NA determined by DeMets and others (1990) from ridge-transform data in the Cayman Trough and Swan-Oriente faults but excluding trench slip vectors. The approximate uniformity of Ca-NA over the last 25 m.y. (Rosencrantz and others, 1988) and of NA-SA over the same duration (Pindell and others, 1988) suggests current Ca-Am velocity in the southeastern Caribbean can be regarded as constant through Neogene time.

Southern Lesser Antilles arc system

The southern Lesser Antilles arc system comprises three main divisions (Fig. 2a): the arc platform, the forearc, and the backarc of which the last comprises the Grenada Basin and Aves Ridge.

The arc platform bisects the arc system and extends with convex-east trace from Margarita through Martinique to an intersection with the Greater Antilles in the Virgin Islands. The platform is defined by a combination of relatively shallow water depth, high free-air gravity (Fig. 2c), thin undeformed sediment cover (≤2 km), and locally, active volcanoes. The active volcanic chain, existent since middle Miocene time, surmounts the platform's western or rear margin where the platform trends N ± 30° (Fig. 2).

Foward (east and south) of the arc platform is the Lesser Antilles forearc, which consists of a series of forearc basins and the extensive Barbados accretionary prism (Fig. 2a). The forearc basins (Carupano Basin, Tobago Trough, Lesser Antilles Trough) contain little-deformed stratal sequences that are as thick as 12 km (Fig. 2b) and probably at least as old as Eocene at the lowest levels (Pereira and others, 1985; Speed and others, 1989). In the Tobago Trough, such sequences thin toward the arc platform on its outer flank, principally due to arcward pinchouts of higher strata that are mainly of post–early Miocene age. The pinchouts imply Neogene uplift of the arc platform relative to the Tobago Trough, which took place apparently by eastward downtilting of the arc platform with little faulting on the tilted flank.

The accretionary prism lies between inner and outer deformation fronts (Fig. 2b) at which forearc basin strata and Atlantic sea-floor sediments, respectively, have undergone progressive deformation and addition to the tectonic edifice of the accretionary prism. The prism has undergone continuous or episodic growth since at least the Eocene (Torrini and Speed, 1989). We interpret the accretionary prism to extend southwestward to coastal Venezuela where it is deeply unroofed and consists of metasedimentary tectonites and is there called the Paria-Trinidad terrane (Fig. 2a; Russo and Speed, 1992). The unroofing is thought to have been caused by the collisional overriding of South America by the western reaches of the southern Lesser Antilles forearc (Speed, 1985).

The Tobago terrane is named for basement rocks that occur within the forearc (Fig. 2a; Speed and others, 1989). From exposures on Tobago (Maxwell, 1948; Rowley and Roobol, 1978; Frost and Snoke, 1989; Snoke and others, 1990) and along the coast of the Paria Peninsula of eastern Venezuela (Bladier, 1977; R. Speed, unpublished data), and from well data in the Carupano Basin (Bellizzia, 1985; Pereira and others, 1985), the basement of the Tobago terrane is thought to consist of probably modified oceanic crust overlain by island-arc magmatic complexes and by sediments that include Early and Late Cretaceous ages. The apparently upramped southern edge of the Tobago terrane can be tracked by short wavelength magnetic anomalies (Speed and others, 1984). The western edge has an uncertain locus in the submarine region east of Margarita because rocks of the Tobago terrane do not exist in Margarita (Ave Lallement and Guth, 1991). The Tobago terrane is depositional basement to Cenozoic basinal strata of the southern Tobago Trough and the Carupano Basin (Fig. 2a; Speed and others, 1989). It is a question whether the Tobago terrane is an isolated crystalline fragment within the accretionary forearc or is continuous northwest below the arc system with crust of the central Caribbean Plate (Venezuelan Basin, Fig. 2a).

In the backarc region are the Grenada Basin and the Aves Ridge (Fig. 2a; Bunce and others, 1970; Kearey, 1974; Westbrook, 1975; Bowin, 1976; Biju-Duval and others, 1978; Boynton and others, 1979; Speed and others, 1984; Bouysse and others, 1985; Pinet and others, 1985; Bouysse, 1988). The Grenada Basin is floored apparently by oceanic crust in its deep southern half (south of 14° N) (Boynton and others, 1979) and crust of uncertain origin in its shallower northern half. Magnetic anomalies over the deep, apparently oceanic part of the Grenada Basin provide no credible indication of spreading orientation or age

(G. K. Westbrook in Speed and others, 1984). Speed and Walker (1991) proposed that oceanic Grenada Basin crust extends eastward across the southern arc platform (Fig. 3) and is greatly uplifted and exposed as pillow basalt in the Grenadines. Such basalt is middle Eocene and of spreading origin, implying that formation of the Grenada Basin was partly or wholly Eocene, an age in consonance with the Basin's bathymetry and heat flow (Boynton and others, 1979).

The southern margin of the Grenada Basin is a deformation belt (SGBDF, Fig. 2b) marked by north-verging thrust or reverse faults that brought the arc platform above southern Grenada Basin crust in Miocene or earlier time (Pinet and others, 1985). The eastern margin of the Grenada Basin and contact with the arc platform from Grenada north is a zone of steeply dipping faults, which is discussed below in detail.

The Aves Ridge is a submerged bathymetric rise that trends north-south between the South American shelf and the northern end of the Lesser Antilles arc. It lies between the oceanic lithosphere of the Venezuelan Basin on the west (Fig. 2a) and the oceanic Grenada Basin on the east. The Aves Ridge probably emerged as an island arc on Caribbean (Venezuelan Basin) crust in early Paleogene time and possibly before (Malfait and Dinkelman, 1972; Kearey, 1974; Bouysse, 1988).

The Paleogene convergent margin was probably on the Atlantic side of the Aves Ridge but the orientation, north or west facing, of the slab is unknown. Extinction of the arc magmatism of the Aves Ridge was probably concurrent with the Eocene opening of the Grenada Basin, which left the Aves Ridge as a remnant arc. An active frontal Paleogene magmatic arc probably then evolved on the Atlantic side of the Grenada Basin (Kearey, 1974; Boynton and others, 1979; Speed and Walker, 1991). An objective of this study was to find the whereabouts of the frontal Paleogene arc.

The crest of the Aves Ridge is topographically rough and includes steepwalled linear and equant uplifts that expose pre-Neogene rock on their flanks (Fox and Heezen, 1975) and cause local positive free-air gravity and short wavelength magnetic anomalies (Kearey, 1974; Speed and others, 1984). Heat-flow measurements (Clark and others, 1978) on the ridgecrest imply locally high temperature gradients, as great as those in the active Lesser Antilles arc. Such properties suggest active extensional horst-graben structure affects the Aves Ridge, and the abundant normal faults of the northern Grenada Basin (Fig. 2b; Bouysse, 1988) may be correlative structures. Therefore, contemporary regional horizontal extension probably affects both the Aves

Figure 3. Map showing possible distribution of low V_p (no pattern) and high V_p (patterns) basements in southeastern Caribbean. Refraction profiles x–x' and z–z' from Boynton and others (1979); profile y–y' modified from Officer and others (1959). On x–x', alternative depths of 6.9 km/sec refractor shown: dashed line is minimum depth, solid line is mean depth, from Boynton and others (1979). SGBDF is southern Grenada Basin deformation front.

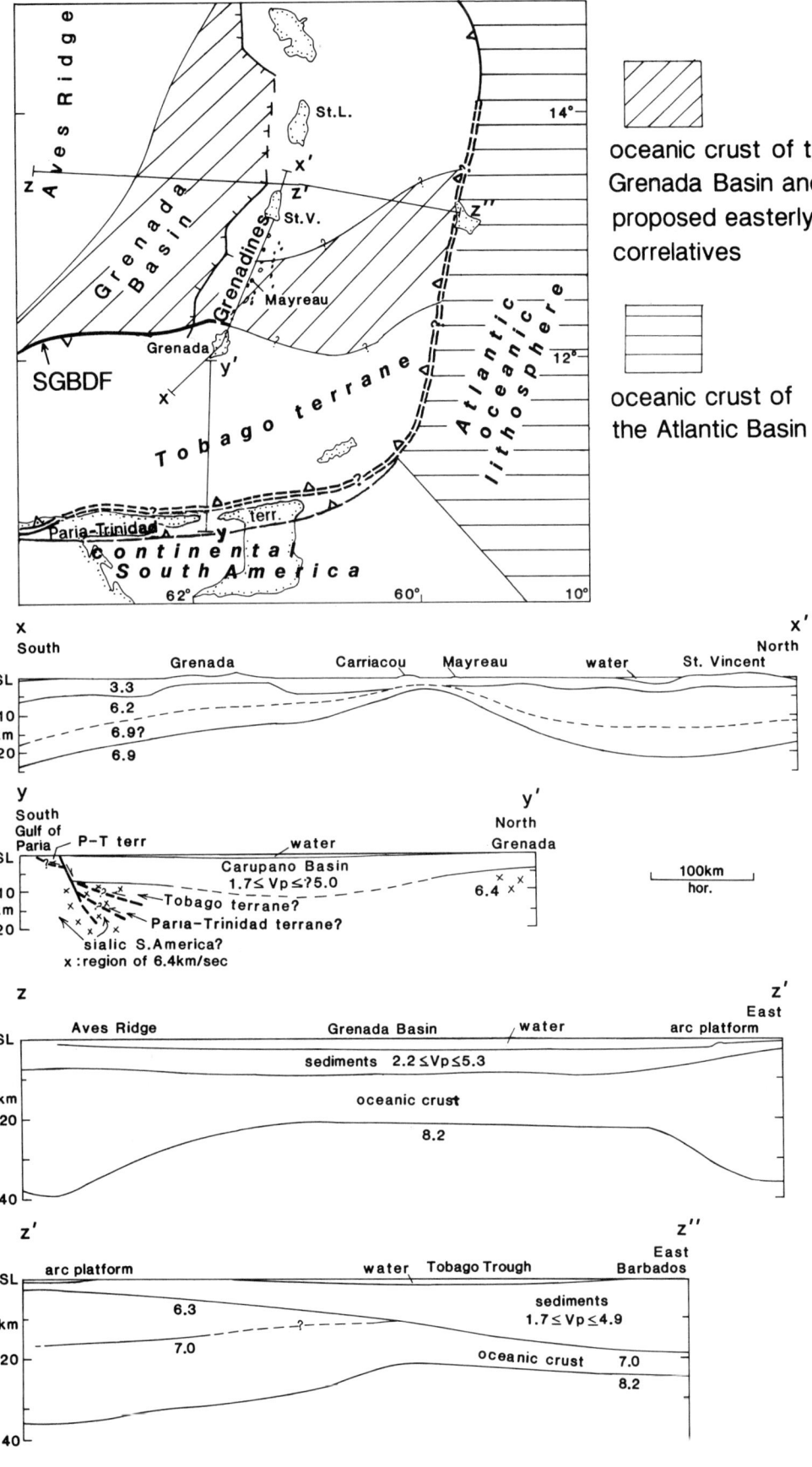

Ridge and northern Grenada Basin. Such extension began in Miocene time and is superposed on earlier structures related to the backarc development of the Grenada Basin and Aves Ridge in Eocene and possibly earlier times.

STRUCTURE OF THE ARC PLATFORM

Morphology

The southern Lesser Antilles arc platform (SLAAP) is a flattopped ridge (Figs. 1, 2), 30 to 50 km wide, that is ≤200 m below sea level except at islands and at marine passages north and south of St. Vincent. The platform extends from Margarita on the southwest in an east-convex trace north to an intersection with the Greater Antilles, as tracked by bathymetry and by linear free air gravity (Fig. 2c) and magnetic highs (Westbrook, 1975; Bowin, 1976; Speed and others, 1984). The western flank of SLAAP descends steeply, 5 to 15° and commonly 10° at northwest-facing slopes, to the deep (3 km), flat-floored Grenada Basin (Figs. 1, 2). The eastern flank of SLAAP descends gradually (about 1.5°) into the Tobago Trough, which is bowl shaped and has a maximum water depth of 2.5 km. Sediment thicknesses in the Grenada Basin are 7 to 12 km, increasing from north to south (Speed and others, 1984), and up to 12 km in the Tobago Trough, generally increasing west to east to an inner deformation front (IDF, Fig. 2b; Speed and others, 1989). The southern flank of SLAAP is buried by Neogene sediments of the Carupano Basin whose strata are as thick as 15 km (Pereira and others, 1985; Speed and others, 1989).

Crustal structure

We discuss the deep structure of SLAAP with views toward the uplift and tectonic history of the platform and, specifically, whether it is composed mainly of old crystalline basement or of Neogene volcanigenic rock. The data employed are chiefly highly generalized refraction velocity-depth models, which are reversed and use first arrivals (Boynton and others, 1979) and gravity (Westbrook, 1975; Kearey and others, 1975; Bowin, 1976; G. Westbrook in Speed and others, 1984). Seismic reflection sections indicate that crystalline basement and nonreflective bodies shallowly underlie the platform, as discussed in the next section.

Refraction models (Fig. 3) imply that the SLAAP is underlain shallowly (less than a few kilometers) by low- and high-velocity crusts, respectively, 6 to 6.5 and 6.9 to 7 km/sec P-wave velocities. The high-velocity crust directly or nearly underlies the sedimentary layer in the region of the central Grenadines, and it dips south and north below opposing wedges of low-velocity crust (Fig. 3, section X–X'). The existence of shallow high velocity crust below the central Grenadines is corroborated by a free-air gravity maximum (Fig. 2c) over a region whose average topography is close to sea level. In contrast, free-air-gravity maxima over the high islands of St. Vincent and St. Lucia are probably mainly topographic effects. A Moho reflector has not been detected in the SLAAP south of St. Vincent. The southern wedge of low-velocity crust thickens south of the SLAAP and probably is partly the Tobago terrane (Speed and Walker, 1991).

The nature and origin of the shallow high velocity basement in the Grenadines are important constraints on the evolution of the SLAAP. Three alternative hypotheses are (1) lower arc crust that extends the full length of the arc platform but which is arched and unroofed in the Grenadines, (2) old oceanic crust on which the Neogene arc has been built but which has had either anomalously little development of arc crust or has been unroofed as in 1, and (3) relatively young oceanic crust that developed between older terranes with low-velocity upper crusts. Speed and Walker (1991) argued that an oceanic rather than lower-arc crustal origin is more likely and that young oceanic crust (hypothesis 3) is the more probable. The latter deduction comes from the recognition that Eocene pillow basalts of spreading origin, the oldest exposed rock units in the Grenadines (Mayreau Basalt, Cherry Hill Basalt), may belong to or be thrust up from the high velocity crust. We consider, therefore, that the shallow basement of the central Grenadines is oceanic crust that is wholly or partly of mid-Eocene age. The northern and southern boundaries of this basement may be either rift-drift margins, which accounts for the wedge shapes of adjacent low-velocity crusts, or post-Eocene faults, the southern of which may be an easterly continuation of the SGBDF that lies between the arc platform and the oceanic southern Grenada Basin (Figs. 2b, 3).

The oceanic crusts thought to underlie the southern Grenada Basin and the central Tobago Trough (Edgar and others, 1971; Westbrook, 1975; Boynton and others, 1979; Speed and Walker, 1991) are each at least as old as Eocene (Boynton and others, 1979; Speed and others, 1989). An approximate subcrop distribution of oceanic and nonoceanic crystalline basement in the southeastern Caribbean can be drawn as shown in the map in Figure 3. It is reasonable to consider that the shallow oceanic crust of the central Grenadines is or was continuous with and originated with oceanic crusts of adjacent basins.

To conclude, data and hypotheses point to the existence of shallow crystalline basement of Paleogene age in the SLAAP. This basement was uplifted in the SLAAP relative to correlatives in modern basins to the east and west. Although the SLAAP basement is intruded and partly covered by rock of the Neogene arc, the volume of Neogene rock may be subordinate, at least south of St. Vincent. Moreover, the SLAAP is not simply an edifice of arc-derived extrusive rock with a central intrusive vent system.

Flank structures

We address here the mechanisms, kinematics, and timing of the uplift of the basement-cored SLAAP with respect to the probably equivalent basements of the adjacent Grenada Basin and Tobago Trough. We employ information taken from seismic reflection sections in a study by P. L. Smith and R. Speed (in preparation). There are evident major differences in the western and eastern flank structures of the SLAAP between Grenada and St. Vincent, and these permit the SLAAP to be interpreted as a half horst with steeply faulted western flank and a gently tilted, east-down flank. The amplitude of relief of basement across the SLAAP to each side may be as great as 12 km.

The western flank of the SLAAP (Fig. 4A to 4C) has characteristically steep slopes, commonly 10 to 15° where the flank is smooth and contains no surficial reflectors (Fig. 4B). By their content of old rock (lower Pliocene and Eocene) dredge hauls suggest such steep, smooth slopes are retreating (Westercamp and others, 1985). The steep slopes may be bedrock surfaces. Locally the western flank is more shallowly dipping, 4 to 5°, and is a rough surface underlain by sediment or rock yielding a field of diffractions. Here, we interpret the slope to be surmounted by a series of disturbed, slumped masses (Fig. 4C). From the base of slope westward in the Grenada Basin, subhorizontal reflectors can be detected below the seabed for about 3 sec two-way time (TWT; Fig. 4A, B). These are slightly upturned at the base of slope and eastward are lost in a zone of incoherence and diffractions that characterizes the flank and interior of the SLAAP. The strata imaged at the basin margin have different reflection characteristics upsection. Below, strata are mainly tabular and conformable whereas upsection, mounded structures and detachments with local discordance exist. In Figure 4A, a prominent shallow detachment surfaces at the base of slope. There are no data with which to date reflectors at the change in reflector style.

Our seismic reflection study (P. L. Smith and R. Speed, in preparation) provides no direct indication of the mechanism or age of formation of the steep western flank of the SLAAP. We propose, however, that it is a zone of normal faulting, as did Pinet and others (1985) and Bouysse (1988) without explanation, and that the faulting probably began in the Miocene or perhaps slightly before. Our evidence for normal faulting is the following: (1) lower Miocene to middle Eocene basinal sedimentary rocks that crop out on the arc platform prove its uplift relative to sea level and imply its uplift relative to basement in the Grenada Basin, which is probably correlative with shallow basement of the platform; (2) the upward appearance of mounded structures in the Grenada Basin margin suggests the onset of slumping and (or) fan deposition and the formation of a slope to the east; and (3) the seismically incoherent zone below the slope is permissible for a zone of steeply dipping normal faults. Alternatively, if the relative uplift were a product of folding, at least part of a limb should be recognized in the seismic sections, and if the scarp were the front of an imbricate thrust system, Grenada Basin strata would be expected to dip down below the thrust front. Such structures are not evident.

Our assignment of age of normal faulting derives from ages of strata on the arc platform that record the transition from basinal to shoalwater conditions of deposition (early and middle Miocene) and the age of beginning of tilting on the east flank. The onset of uplift of the SLAAP may have preceded this.

The eastern flank of the SLAAP (Fig. 4C, D) has shallow slopes (1 to 2°) and generally thick sediment cover, in marked contrast to the western flank. In the subsurface, the eastern flank consists of basement and older basinal sediment cover that is tilted relatively steeply to the east, perhaps as much as 20°, from a hinge with the subhorizontal (but high relief) basement floor of the central Tobago Trough (Figs. 4C, D). The tilted older strata are generally undisrupted (Fig. 4D). Minor faulting may have occurred near the updip end of the tilted segment. Younger strata progressively onlap the tilted strata or basement (Figs. 4C, D) and include minor unconformities and facies changes to more mounded deposits and (or) slumps near the upper slope break (Fig. 4D).

The beginning of tilting on the eastern flank occurred between 0.1 to 0.4 sec TWT before horizon C (Fig. 4C) in the Tobago Trough, which is estimated to be between 16 and 20 Ma (early to middle Miocene) by Speed and others (1989). At sedimentation rates and velocities figured by Speed and others (1989), this suggests that tilting began no earlier than 25 to 23 Ma, latest Oligocene or early Miocene.

To conclude, the strong evidence for tilting and only minor disruption of the SLAAPs eastern flank and that for abrupt large offset of basinal strata and basement across the western flank lead to the interpretation that the SLAAP is a half horst or trapdoor structure that was uplifted relative to sea level and to adjacent basin floors. The onset of normal faulting on the western flank and tilting on the eastern was in early Miocene and possibly latest Oligocene time.

Crestal structure

Seismic coverage across the top of the SLAAP is meager, and most sections available to us show little structure—only horizontal reflectors for 0.2 to 0.5 sec TWT below the seabed and incoherence below them (P. L. Smith and R. Speed, in preparation).

The seismic section of Figure 4C, at the latitude of Carriacou, however, shows slightly deeper reflections. There, an unconformity can be recognized above synformal reflections (Fig. 4C) and below unfaulted horizontal reflections. To the east, the suprajacent reflections are arcuate and indicative of mounded deposits that prograde farther east across the eastern platform margin (mounds I, Fig. 4C). The unconformity can be traced eastward across the margin and through a stack of diffractions into the tilted flank. There, we correlate it with horizons of late Miocene age. Vague horizontal reflectors lie above the reflectors of mounds I (Fig. 4C), and these are surmounted by arcuate reflectors that represent a younger mounded structure (mounds II, Fig. 4C) with a steep east face at today's seabed and a fanlike structure below it. Mounds I and II are readily interpreted as carbonate buildups at the platform edge; the horizontal reflectors to the west represent back-reef deposits. The unconformity below the buildups has subsided about 600 m relative to sea level, assuming a sediment velocity of 2 km/sec. The basinward progradation of mound II relative to mound I suggests the rate of buildup growth exceeded the rate of subsidence.

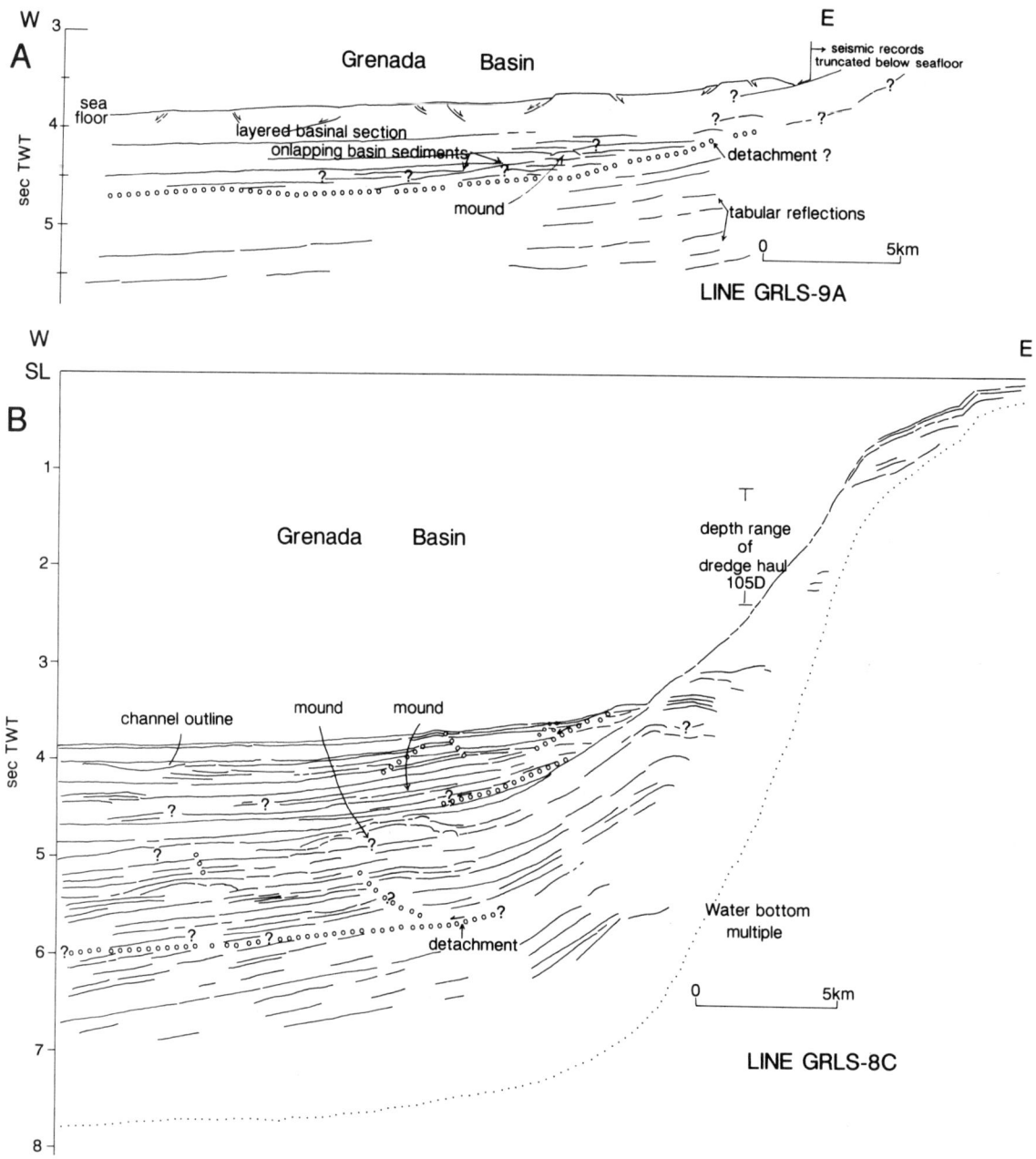

Figure 4. Line drawings of seismic sections crossing flanks of SLAAP; tracks on Figure 1B; A, eastern margin of Grenada Basin west of Bequia; B, western flank of SLAAP west of Mustique; C, western flank and crest of SLAAP south of Carriacou; D, crest and eastern flank of SLAAP south of Carriacou; E, eastern flank of SLAAP east of Grenada; GRLS -8C and -9A are 24 fold; PC9304 and 9306 are 60 fold; all unmigrated.

By virtue of their age and structure, the slightly deformed strata below the unconformity (Fig. 4C) on the platform are likely to be middle Miocene formations exposed on Carriacou. These record the first shoaling recognized in rocks of SLAAP. If the correlation is correct, it is not surprising that older formations of the SLAAP cannot be recognized in seismic sections because their deformation is great enough to preclude coherent reflections.

The crest of the SLAAP has rocks as old as middle Eocene above sea level and probable middle Miocene beds at 600 m below. This implies considerable relief has either persisted since or (and) been created during Neogene time. The absence of faulting of strata above the dated Miocene unconformity suggests the former, but seismic reflection data on the crest are too scanty to be certain.

Neogene magmatism

The emplacement of Neogene magmas into the crust of what is now the southern Lesser Antilles arc platform may have

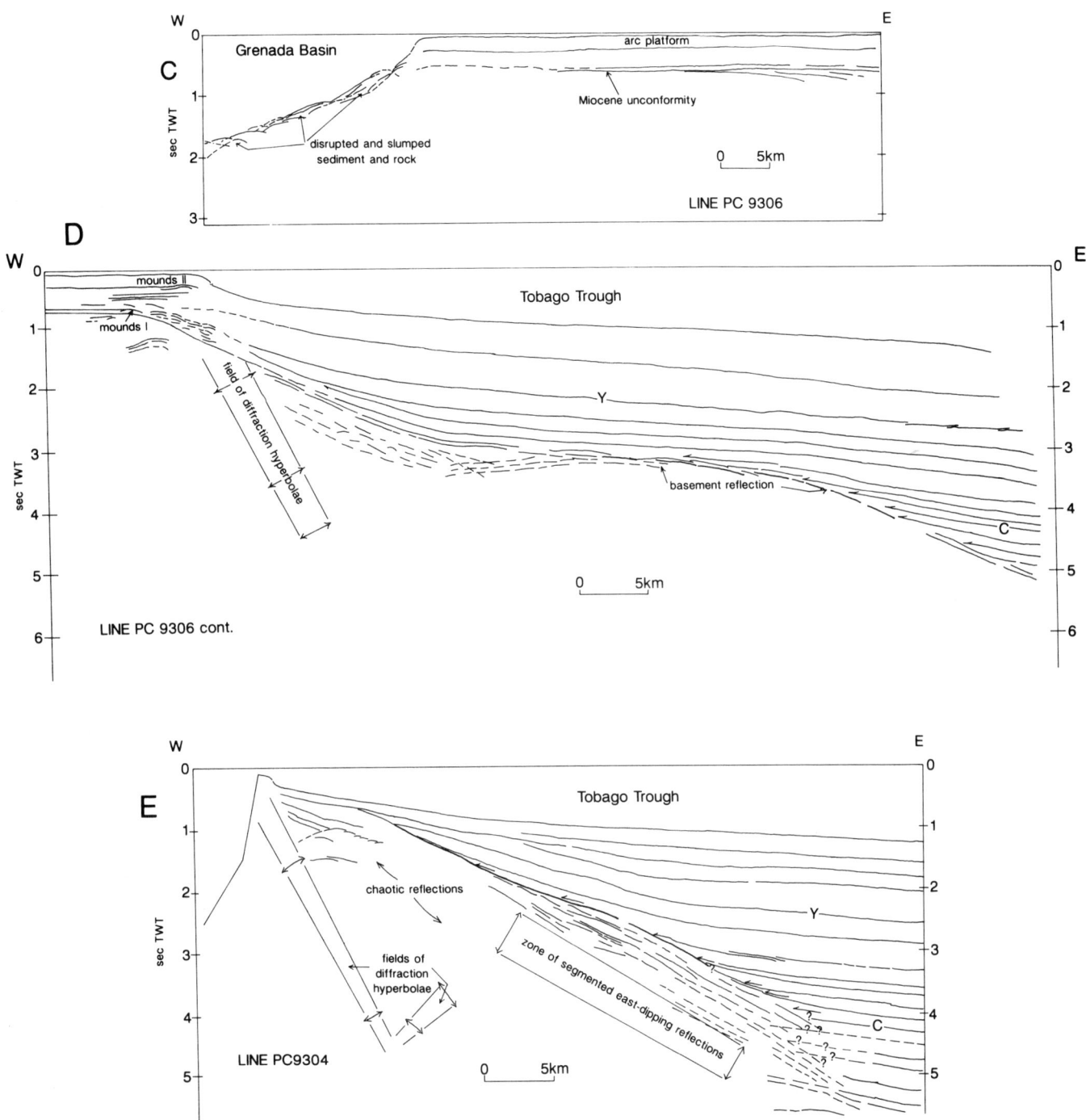

had substantial effects on the evolution of the platform, namely its history of deformation and of uplift relative to sea level and adjacent basins. Below, we discuss what is known about the age range and emplacement of Neogene magmatic rocks in the SLAAP with a view to such effects. We use Neogene in the sense of Palmer (1983) to indicate time from the beginning of the Miocene to the present.

The chain of islands along the western rim of the SLAAP from Grenada north marks the principal locus of Neogene magmatism (Fig. 1A, B). The large islands of the SLAAP (Grenada, St. Vincent, and St. Lucia), are sites of concentrated venting, and each includes ancient and late Quaternary volcanoes (Arculus, 1976; Rowley, 1978; Le Guen de Kerneizon and others, 1983). In contrast, the Grenadines present little evidence for recent volcanic activity except at Kick-Em-Jenny (Fig. 1B), which is currently active. The Grenadines differ from the large islands by their low elevations and higher proportion of exposure of older rocks. The minor volume of Neogene magmatic rocks in the Grena-

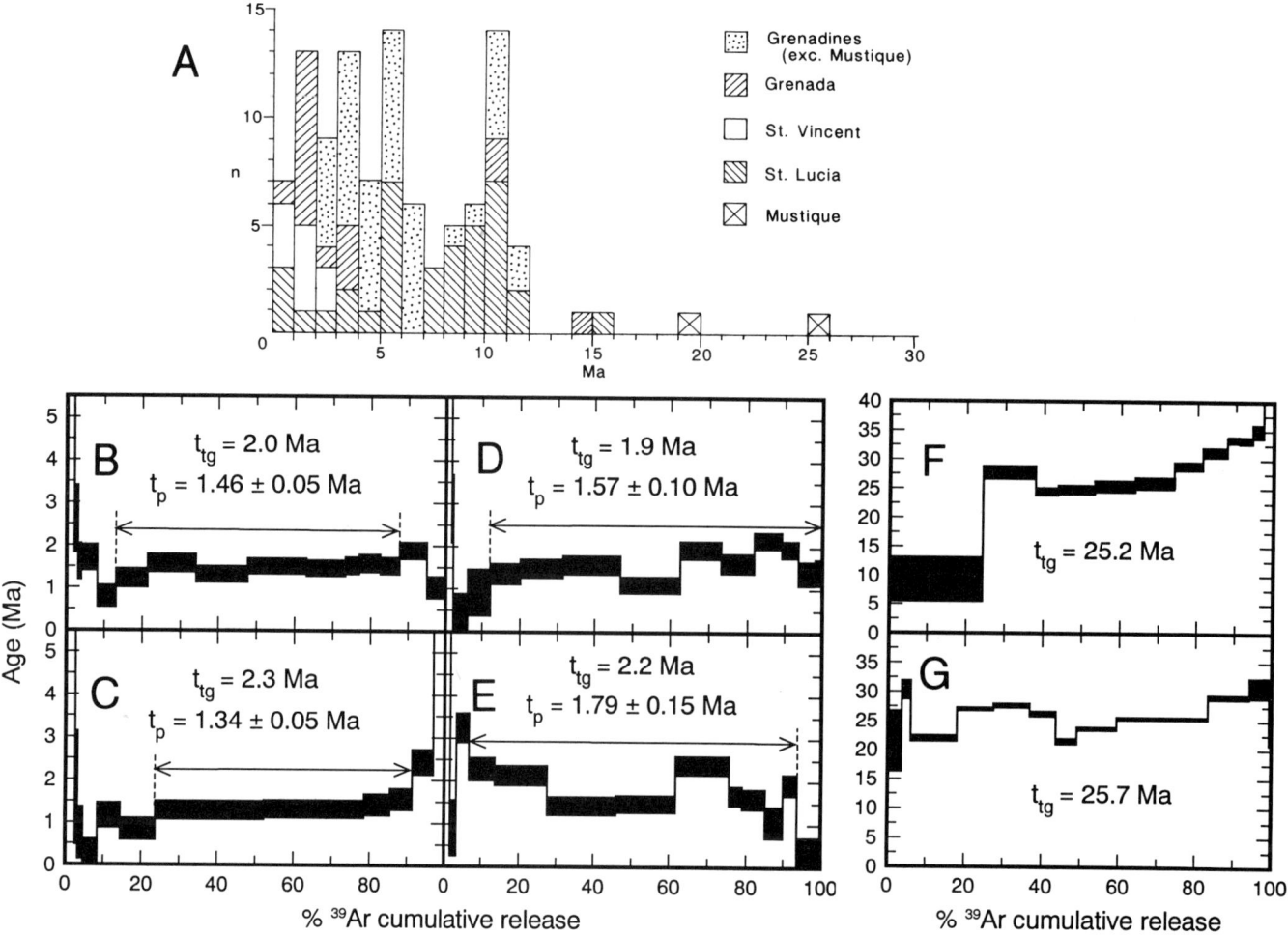

Figure 5. Radioactive age data from Neogene magmatic rocks of the southern Lesser Antilles arc platform. A, K-Ar dates from Briden and others (1979), Le Guen de Kerneizon and others (1983), Westercamp and others (1985), and Table 1; B and C, ^{40}Ar/^{39}Ar spectra for two fractions of hornblende phenocrysts from sample 91-1, andesite dome at Mt. Craven, Grenada (Fig. 6B); D and E, same as B and C, sample 91-4, andesite dome at Mt. Craven, Grenada (Fig. 6B); F, ^{40}Ar/^{39}Ar spectrum of whole rock from mafic dike of eastern Mustique (Fig. 26B), sample 89-4; G, ^{40}Ar/^{39}Ar spectrum for plagioclase phenocrysts—same site and sample as F. t_{tg} is total gas date; t_p is plateau date.

dines, however, may be partly due to erosion because the magmatic rocks there are mainly intrusions.

Figure 5A shows the frequency of 105 K-Ar dates of magmatic rocks from Grenada to St. Lucia, compiled from Briden and others (1979), Le Guen de Kerneizon and others (1983), Westercamp and others (1985), and Table 1 of this paper. Ninety-five percent of these are younger than 12 Ma. Within this duration, Grenada and St. Lucia have broad age distributions, whereas dates on St. Vincent are no older than 3 Ma, and those of the Grenadines are all pre-Quaternary. The young and narrow range of the St. Vincent dates may be due to sampling bias. The span of Grenadine dates, however, supports the morphologic observation that the magmatism was extinguished there sometime ago except for Kick-Em-Jenny.

We dated Neogene magmatic rocks radiometrically from Grenada, Mustique, and Mayreau (Fig. 1, Table 1). The analyses are by K. A. Foland.

On Grenada, hornblende andesite of the Northern Domes center at Mt. Craven (Fig. 6A; Arculus, 1976) was sampled with the objective of evaluating a previous date for this body of 21.2 Ma (hornblende, K-Ar; Briden and others, 1979), by far the oldest apparent age among Neogene magmatic rocks of the SLAAP. Our samples came from two sites (Fig. B), one of which was the source of the specimen dated 21.2 Ma according to directions provided by Richard Arculus. We used the ^{40}Ar/^{39}Ar step-heating technique on two fractions of hornblende from each site (Figs. 5B to E). The four spectra indicate plateaus between 1.2 and 1.9 Ma and total gas dates, which should closely approx-

TABLE 1. ^{40}Ar/^{39}Ar DATES OF IGNEOUS ROCKS FROM GRENADA AND MUSTIQUE AND K-Ar ANALYTICAL DATA AND DATES FOR IGNEOUS ROCKS FROM MUSTIQUE AND MAYREAU*

Island	Sample Number	Dated Material[†]	Plateau Date (Ma)	Total Gas Date (Ma)	K (wt %)	^{40}Ar rad (10^{-11} mol/g)	$\frac{^{40}\text{Ar rad}}{^{40}\text{Ar Total}}$	K-Ar Date (Ma)
Grenada	91-1a	Hbl	1.46 ± 0.05	2.0				
	91-1b	Hbl	1.34 ± 0.05	2.3				
	91-4a	Hbl	1.57 ± 0.10	1.9				
	91-4b	Hbl	1.79 ± 0.15	2.2				
Mustique	89-6	Wr			0.389	1.31	0.101	
					0.384	1.29	0.105	
					0.389			
				Average	0.389	1.30		19.2 ± 0.6
	89-4	Wr			0.497	2.19	0.284	
					0.486	2.16	0.258	
					0.499			
				Average	0.514	2.18		25.2 ± 0.4
	89-4	Wr		25.7				
	89-4	Plag		25.2				
Mayreau	87-022	Wr			0.512	1.20	0.144	
					0.516	1.34	0.155	
					0.516	1.25	0.166	
				Average	0.514	1.26		14.1 ± 0.8
	87-026	Wr			0.753	1.57	0.212	
					0.743	1.86	0.243	
					0.738	1.40	0.194	
				Average	0.745	1.61		12.4 ± 1.8

*Samples sites on Figures 6B, 22, and 26. ^{40}Ar/^{39}Ar spectra in Figure 5B through 5G. Analyses by K. A. Foland. Analytical uncertainty in date is 68% confidence level.
[†]Hbl = hornblende; Wr = whole rock; Plag = plagioclase.

imate K-Ar dates, between 1.9 and 2.3 Ma. The upshot of the new dating, discussed further in the section on Grenada, is that the 21.2-Ma date is invalid as an age of crystallization and that the oldest apparent age in Grenada, also perhaps questionable, is now 14 Ma.

We dated two dikes on Mustique by whole-rock K-Ar (Table 1). From the older of these dikes, we also dated phenocrystic plagioclase and whole rock by ^{40}Ar/^{39}Ar (Table 1; Fig. 5F, G). The objectives were to test our field observations that the igneous bodies are intrusions in the Oligocene wallrocks, not lavas as mapped by Westercamp and others (1985). As discussed in the section on Mustique, our K-Ar and total gas dates are about 19 and 25 Ma for the two dikes, thus supporting the interpretation that igneous activity was after sediment deposition. Because of discordance in the argon spectra, however, the exact age of intrusion of at least the older dike is uncertain. The dates may either record local very late volcanism of the Paleogene arc, a hydrothermal event that adjusted the K-Ar composition of the igneous rocks, or a fortuitous sample of early magmatism of the Neogene arc not yet recognized elsewhere in the SLAAP.

On Mayreau, we dated two samples of Eocene Mayreau Basalt by whole-rock K-Ar (Table 1). The dates, 12 and 14 Ma, may record the age of Neogene dikes that widely intrude Mayreau Basalt.

To conclude, the magmas of the modern-Neogene Lesser Antilles arc reached the surface and near surface of the SLAAP in copious volume between about 12 Ma and the present. Minor volcanism probably occurred as far back as 15 Ma, and it is uncertain whether any Neogene eruption preceded that date.

The form of Neogene intrusions in the Grenadines is highly varied: dikes, sills, plutons, and diatremes. Basalt occupies dikes and sills that are commonly <10 m wide and exceptionally 30 m. More siliceous rock occurs generally in wider, more irregularly shaped dikes or narrow plutons. The only intrusion that is approximately equant in plan is the siliceous pluton of eastern Union Island, which is at least 0.5 km in diameter. Diatremes are either tabular or irregular and are chiefly associated with dike tips and pluton walls.

Siliceous rocks and perhaps basalt may have been concentrated at certain loci of intrusion in the western Grenadines. In Carriacou, we suggest the locus is associated with the crest of a local arch in the older rocks. In Grenada, Neogene extrusions

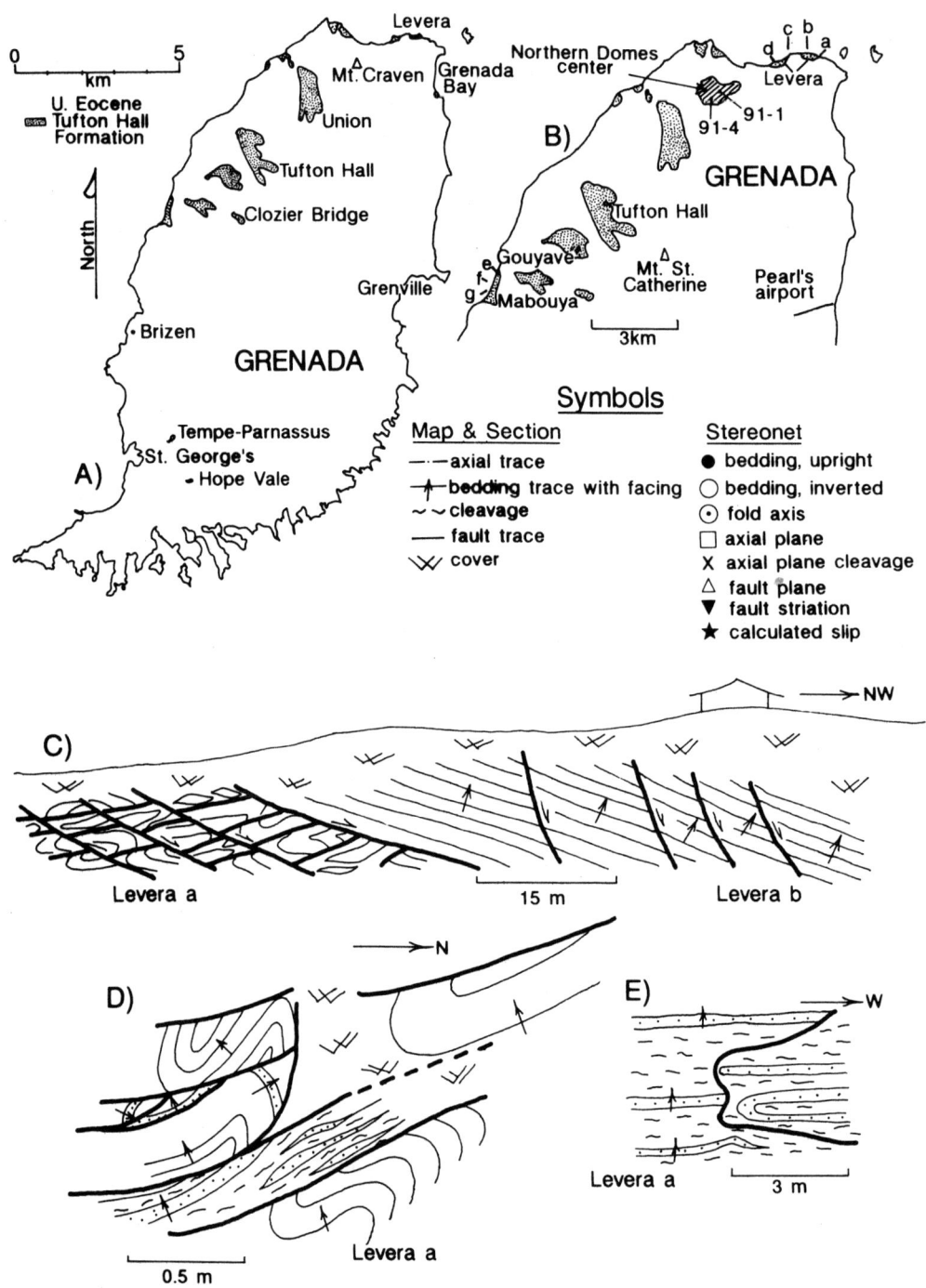

Figure 6. Map, cross sections, and orientation data for Tufton Hall Formation of Grenada. A) Map of Grenada showing areas of outcrop (shaded) of Tufton Hall Formation and sites (Tempe-Parnassus, Hope Vale, and Brizen) where other Paleogene rocks are exposed. B) enlarged map of northern Grenada showing study areas (small letters) referred to in succeeding illustrations and sample sites referred to in Figure 5. C) Cross section showing two structural associations: upright homocline of Levera b and disrupted folded beds of Levera a. D) Cross section at Levera a showing curved faults. E) Cross section at Levera a showing curved fault plane, probably a lateral or oblique ramp. F) and G) Orientation diagrams for disrupted folded rocks in eastern Levera a. H) Cross section illustrating northerly heave at thrust ramp and northerly vergence in footwall fold at Levera c. I) Cross section showing recumbent fold with disrupted hinge at Mabouya f. J) Cross section showing folded fold in local deformation band within a homocline at Mabouya e. K) and L) Orientation diagrams for disrupted folded strata at Levera c and d. M) Cross section of major recumbent fold pair at Mabouya g; hinges of both folds are faulted. N) and O) Orientation data for recumbent folds and disrupted folded strata at Mabouya e and f, P). Orientation data for major recumbent fold at Mabouya g, illustrated in Figure 6M.

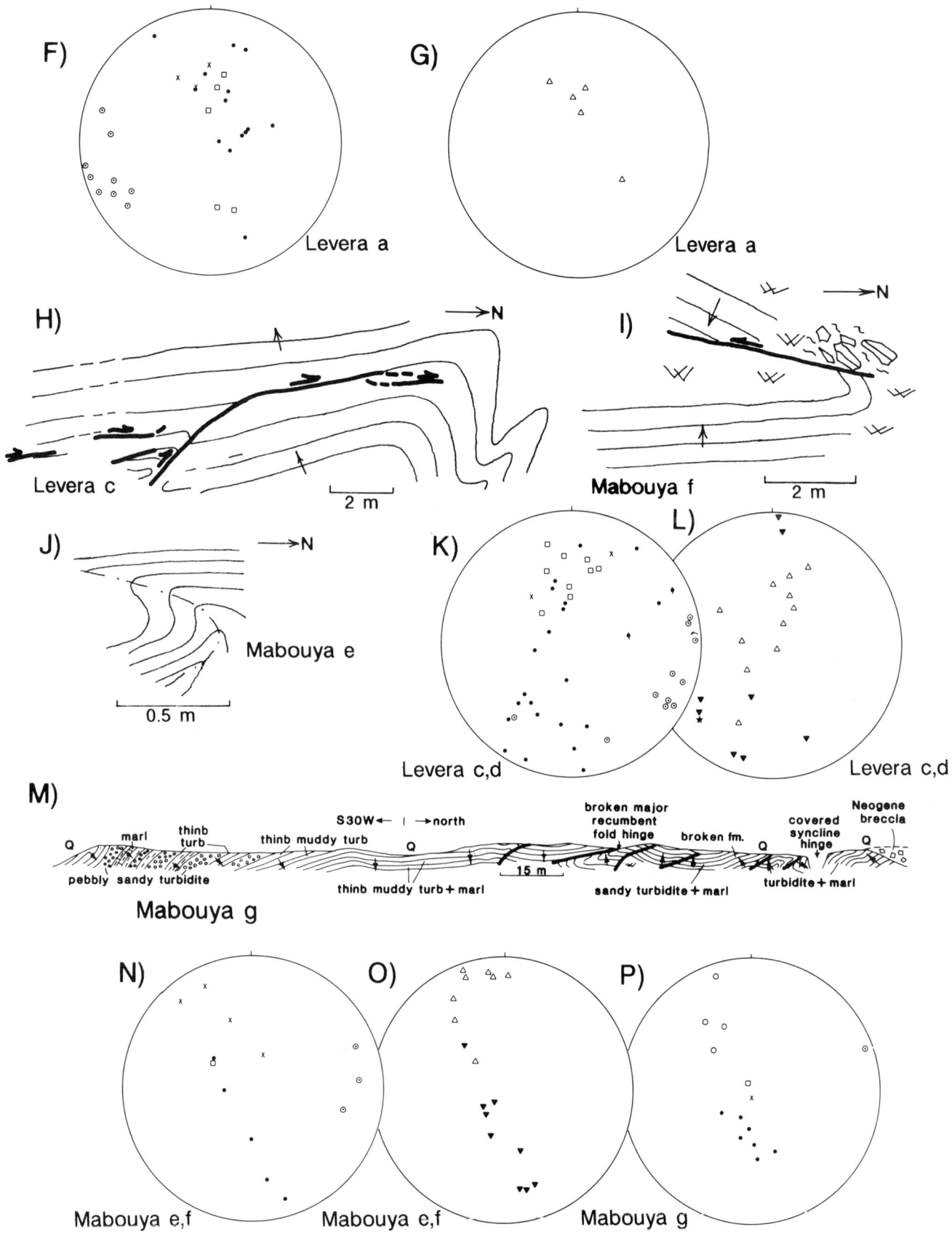

cover most of the island (Arculus, 1976), but no magmatic dikes cut the subjacent older rocks, at least in their sparse outcrop belt along the northern and western coasts (Fig. 6). This suggests that magmas of Grenada have been channeled through central conduits and supports the idea of generally restricted or concentrated intrusion at certain loci.

We have measured orientations of Neogene dikes on Carriacou and Mayreau. As a whole, their distribution of strikes is nearly random, implying that their emplacement was not controlled simply by arc-normal extension. There is one exception in Carriacou of a dike set and normal faults, which both strike approximately north-south. The wide range of strikes of dikes may be related to strains surrounding vertical pipelike magma conduits.

The crustal level at which we observe magmatism in the SLAAP must be at and close to that of the seabed at the onset of Neogene magmatism, as indicated by thermal phenomena (chilled margins of intrusions, meager metamorphism if any, brittle wallrock deformation) and by stratigraphy. It is our impression that the volume addition to the crust at this level is moderate to small (<30%) even in the loci of concentrated intrusion in the western SLAAP. It is possible, however, that large Neogene intrusions exist at depth and have caused substantial volume increase of the SLAAP as a whole. The existence of comagmatic plutonic rocks at depth is indicated by copious cognate inclusions in basaltic lavas in the larger islands (Arculus and Wills, 1980). Nonetheless, it is difficult to prove the size and depth of the chambers from which the inclusions came.

The uplift of the half horst that is the SLAAP was underway early in Miocene time, probably beginning at or before about 23 Ma and completed by about 16 Ma. The emplacement of Neogene magmas in significant volume at levels at and near the present surface started at about 12 Ma, apparently well after emergence of the half horst. Such magmatism is concentrated near the western margin of the half horst, near the normal fault zone at which the southern Grenada Basin broke away and subsided relative to the formerly coplanar crust of the SLAAP. If the uplift of the half horst was caused by or was otherwise related to advection of magmas of the Neogene arc, the early phases of magmatism were wholly or mainly intrusive, and the transit of magma to the surface of the horst took at least 9 m.y.

GEOLOGY OF THE ISLANDS

Introduction

This section presents data on the geology of exposed tracts of older rocks of the SLAAP and gives interpretations of the origins of local features. The presentation is island by island, from south to north. The information content of each island is proportional to the area of exposure of older rocks; thus, Carriacou is given a lengthy treatment, whereas tiny islets, such as Jamesby, are described more briefly. The section following this one synthesizes the geology of the islands with views toward the stratigraphy, magmatism, depositional environments, and deformation history of the SLAAP as a whole. There, Table 4 summarizes properties of all older rock units, and Figure 27 displays their age ranges.

Exact dating of rock units is a difficult endeavor. First, the physical stratigraphy of most units is uncertain, owing to deformation and spotty exposure. The possibility of repetitions and omissions of section by layer-parallel faults exists in all units. The age range determined for one bed, therefore, can rarely be narrowed by age ranges of higher or lower layers except where exposure is exceptionally good. Second, most of the units contain turbidites, and the resedimented fossils in these provide only a maximum age of deposition. True ages of deposition can be obtained from pelagic interbeds and pelagic tops to turbidites, but such layers constitute only a small fraction of some units. Third, the coexisting microfossils are commonly long-lived (>10 m.y.) species, and where these have long durations of overlap, dating is poorly resolved. These uncertainties are particularly limiting in attempts to establish ages of the oldest and youngest beds of a unit and to demonstrate the existence of either internal temporal continuity or hiatuses within or between units.

To express our dating and that of others most exactly, we use a fourfold nomenclature of age ranges—zonal, composite, permissible, and stratigraphic—explained as follows.

We use zonal for the age of a specimen (that is, a hand sample) when the overlap of two or more species is limited to a duration of one or two microfossil zones. A zonal age implies the age of the specimen is known within about 0.5 to 4.0 m.y. In the ideal, a rock unit's age range is the range of a succession of zonal ages.

Composite is used for the age range of a rock unit for which prior workers reported a species list for the unit but not by bed or specimen. In some cases, we have pooled our identifications with theirs. The justifications for using a composite age are that the units so dated are thin, suggesting they may not represent a long duration, and that the ages of pooled species all overlap. The actual age range of a unit could be greater than the composite age.

A permissible age range is used for specimens that do not yield zonal ages but have one or more microfossils that provide some age limit. A permissible range for a unit is the duration between the oldest and youngest limits of the permissible ranges of specimens from the unit. A unit's permissible range suggests the maximum duration a unit may have from what is known of its microfossil content.

The stratigraphic range places age bounds on a unit from contacts with other dated units. The stratigraphic range may be longer or shorter than the unit's permissible range.

We use the nannofloral zonation of Martini (1971) and the planktic foraminiferal zonation of Blow (1969). Correlations used between these two zonations and between them and the geologic time scale are those of Berggren and others (1985). The concept of foraminiferal species used by Saunders in this study follows that in Bolli and others (1985).

Grenada

Introduction. Grenada is extensively covered by Neogene volcanic rocks (Arculus, 1976), which are mainly ≤12 Ma according to 18 radiometric dates (Fig. 5A). Among older rocks of Grenada (Fig. 6), the principal unit and focus of our study is the Tufton Hall Formation, which is exposed along the northern and northwestern coasts. Other older rock units, sparingly exposed, occur at Tempe-Parnassus and Hope Vale, and limestone blocks occur in tuff at Brizen (Fig. 6). Literature concerning older rocks of Grenada includes Martin-Kaye (1958, 1969), Saunders and others (1985a, b), and Speed and Larue (1985).

Tufton Hall Formation. This formation consists of turbidites and minor hemipelagic interbeds of Eocene age (Fig. 6). The turbidites are thin muddy and sandy and thick sandy and pebbly beds of arc-derived volcanic rock fragments, feldspar, and quartz and of skeletal and intraclastic debris. The thin-bedded successions are turbidites, interpreted as of outer fan or basin plain origin, whereas the thicker, coarser beds are channel fills. The skeletal component implies the volcanic source was shallow marine if not subaerial, and the freshness of the volcanic particles suggests coeval extrusion and resedimentation (Saunders and others, 1985a, b; Speed and Larue, 1985).

Dating. The turbidites include hemipelagic tops and interbeds. These contain planktic foraminifers (Table 2) whose composite age is within the *Turborotalia cerroazulensis* zone, P16, late late Eocene. This age was found at all the major outcrop areas of Eocene turbidite on Grenada (Saunders and others, 1985a, b). There is, however, no stratigraphy established among Eocene turbidites, owing to structural complexity. The thickness of the late late Eocene succession, therefore, is unknown.

Structures. All the Eocene turbidite beds are thoroughly faulted, probably in two phases, and are variably affected by folds and cleavage. Structural associations within fault-bounded packets of Eocene turbidite are of three types: mainly homoclinal successions, major recumbent folds, and disrupted folded beds (Speed and Larue, 1985). The three types are now discussed.

<u>Homoclinal successions.</u> Sequences of upright beds as thick as 25 m occur in areas Levera b and Mabouya e (Fig. 6B) and at Tufton Hall (Fig. 6A). Such beds are variably affected by low-angle to bedding-parallel faults, some of which are associated with sporadic disharmonic and folded (Fig. 6J) minor folds. Fault-plane striations and fault-fold relations indicate updip slip. Homoclinal beds are generally uncleaved. They are cut by normal faults that are younger and at a higher angle to bedding. There is no conspicuous difference in layering or lithic character of homoclinal successions from beds of other structural associations. We are unsure whether homoclines evolved as initially detached and unfolded successions of beds or whether they are upright limbs of major folds whose hinges are unexposed or faulted off.

<u>Recumbent folds.</u> A major recumbent fold and related minor folds are in sea cliffs near Mabouya (area g, Fig. 6M). The upright northern limb is exposed over 60 m horizontal length and the overturned southern limb over 100 m between limits of younger cover. Both limbs are cut by south-dipping thrusts of small displacement, and one of these truncates the axial plane of the major fold. The major fold has a subhorizontal axial plane and axis (Fig. 6P) and spaced axial planar cleavage in muddy beds but not in sandy ones. Cleavage is better developed in the overturned than in the upright limb. The fold is tight (interlimb angle <70°), and layer styles are subparallel for sandy beds and nearly similar for muddy ones. The vergence of the major fold near Mabouya is not directly observable but is likely to be northerly because the fold is south closing and because associated faults have northerly heave. Homoclinal successions of Eocene turbidites of Grenada may have evolved as upright limbs to folds of this type.

Bedding-parallel slip occurred during recumbent folding, as indicated by local discordances and by striated bedding surfaces. Such slip was taken up more continuously within cleaved muddy layers where cleavage throughout the layer is striated and slickensided. The cleaved mudstones thus acted as shear zones during low-angle faulting, and the cleavage became scaly. Attitudes of low-angle faults are girdled with bedding, and slip directions are mainly normal to the girdle axis (Fig. 6O). At some places, the low-angle faults are clearly late with respect to major folding, but at others, such faults could have been early and have accommodated layer-parallel shear before and (or) during folding.

High-angle normal faults cut the folded rocks as discrete surfaces in sandy beds but as braided zones of distributed shear, 1 to 10 cm wide, in some intervening mudstones. These late shear zones contain a spaced cleavage that may be superposed on an earlier axial plane cleavage. The high-angle faults are generally east-striking and have dip slip.

<u>Disrupted folded beds.</u> Numerous minor folds of beds and the disruption of hinges and limbs define this third structural association, best exposed at areas Levera a and Levera c (Fig. 6B). The folds are tight to involute, 0.3 to 2 m in minimum width, and of parallel style in sandy beds. Fold axes are shallowly plunging and moderately homoaxial, and axial planes are mainly south-dipping but partly girdled about the mean axis (Fig. 6F, K). Muddy beds include spaced axial plane cleavage and suggestions of an earlier bedding-parallel cleavage in some hinge regions.

Folds are broken by faults of two sets: an earlier with mainly south-dipping closely spaced thrusts, and a later with mainly north-dipping, low-angle normal faults, as shown diagrammatically in Figure 6C. The earlier faults are largely responsible for the disruption. They have widely varied dip angles and are abundantly branched (Fig. 6D, K); some are strongly curved and evidently represent lateral or oblique ramps (Fig. 6E). These faults cut folds in their overturned limb such that intact limbs and elongate blocks within the disrupted folded rocks are generally upright. Some packets in zones of disruption are composed of unfolded phacoids of upright sandstone in a matrix of mudstone whose cleavage is more scaly and striated than that in packets with tabular folded sandstone (Fig. 6D). The phacoid-bearing

TABLE 2. PLANKTIC FORAMINIFERS IDENTIFIED IN OLDER ROCKS OF THE GRENADINES AND GRENADA

Island	Unit	Observer, Ref.*	Sample	Globigerinatheka sp.	G. subconglobata gr.	G. cf kugleri	Morozovella sp.	M. aragonensis	M. bulbrooki	M. spinulosa-lehneri gr.	M. spinulosa	M. lehneri	M. broedermanni	Planorotalites psuedoscitula	Psuedohastingerina micra
Mayreau	Anse Bandeau Formation	JBS, 1990	ts, composite	r	-	-	x	-	-	-	-	-	-	-	-
		PA, 1990	ts, composite	-	x	-	-	x	-	-	x	-	-	-	-
		W, 1985	ts, composite	x	-	-	-	x	x	-	x	-	-	-	x
Jamesby	Marly bed	W, 1985	composite	-	-	x	-	-	-	x	-	-	x	x	x
Baradel	Chert unit	JBS, 1990	ts, composite	x?	-	-	-	x?	-	-	x?	x?	-	-	-
		PA, 1990	ts, composite	-	-	-	-	x	-	-	-	-	-	-	-
		W, 1985	composite, ?method	x	-	-	-	x	x	-	x	x	x	-	-
Carriacou	Cherry Hill Basalt	JBS, 1990	90-46a, site 14	-	r	-	-	-	-	-	-	x?	-	-	-
			90-46b, ts, site 14	-	-	-	-	-	-	-	x?	x?	-	-	-
	Bogles Limestone	JBS, 1990	90-8b, ts, site 27	x	-	-	-	x?	-	-	-	-	-	-	-
			90-34, ts, site 28	-	-	-	-	-	-	-	-	-	-	x?	-
	Belvedere Formation	JBS, 1990	10-9-7-m2, site 29	-	-	-	-	-	-	-	-	-	-	-	-
			90-47, site 20	-	-	-	-	-	-	-	-	-	-	-	-
			90-53, site 24	-	-	-	-	-	-	-	-	-	-	-	-
		W, 1985	Cari 15	-	-	-	-	-	-	-	-	-	-	-	-
			Cari 16, 17 (vic. Dover, site uncertain)	-	-	-	-	-	-	-	-	-	-	-	-
		RJ, 1972	10815, 10816	-	-	-	-	-	-	-	-	-	-	-	-
	Anse La Roche Formation	JBS, 1990	90-28, site 4	-	-	-	-	-	-	-	-	-	-	-	-
		W, 1985	composite, (vic. Gun Point)	-	-	-	-	-	-	-	-	-	-	-	-
		RJ, 1972	10764, ts	-	-	-	-	-	-	-	-	-	-	-	-
	Belmont Formation B	RJ, 1972	10740	-	-	-	-	-	-	-	-	-	-	-	-
			10745	-	-	-	-	-	-	-	-	-	-	-	-
		W, 1985	Cari 19	-	-	-	-	-	-	-	-	-	-	-	-
	Kendeace Formation	RJ, 1972	10788, 10790	-	-	-	-	-	-	-	-	-	-	-	-
			10779	-	-	-	-	-	-	-	-	-	-	-	-
			10817	-	-	-	-	-	-	-	-	-	-	-	-
			10699	-	-	-	-	-	-	-	-	-	-	-	-
			10737	-	-	-	-	-	-	-	-	-	-	-	-
	Carriacou Formation type section, basal echinoidal limestone	RJ, 1972	10754, 10777, 10778, 10787, all composite samples	-	-	-	-	-	-	-	-	-	-	-	-
	unit above algal limestone	RJ, 1972	no sample numbers	-	-	-	-	-	-	-	-	-	-	-	-
Canouan	Canouan Formation	W, 1985	composite, approx. 10-m section	-	-	-	-	-	-	-	-	-	-	-	-
Mustique	Marls at Black Sand Bay and Gallicaux Bay	W, 1985	composite of 8 samples	-	-	-	-	-	-	-	-	-	-	-	-
Union	Sandstone-chert unit	JBS, 1990	ts, composite	x	-	-	-	-	-	-	-	-	-	-	-
Grenada	Tufton Hall Formation	S, 1985a	composite, at type section	-	-	-	-	-	-	-	-	-	-	-	x

TABLE 2. PLANKTIC FORAMINIFERS IDENTIFIED IN OLDER ROCKS OF THE GRENADINES AND GRENADA

Island	Unit	Observer, Ref.*	Sample	Acarinina bulbrooki-spinuloinflata gr.	A. spinuloinflata	Hantkenina sp.	H. mexicana	H. alabamensis	H. primitiva	Truncanorotaloides rohri	T. topilensis	Turborotalia cerroazuelensis gr.	T. pomeroli	T. possagonensis
Mayreau	Anse Bandeau Formation	JBS, 1990	ts, composite	-	-	-	-	-	-	-	-	-	-	-
		PA, 1990	ts, composite	x	-	-	-	-	-	-	-	-	-	x
		W, 1985	ts, composite	-	-	-	-	-	-	-	-	-	-	-
Jamesby	Marly bed	W, 1985	composite	-	-	-	-	-	-	-	x	-	x	-
Baradel	Chert unit	JBS, 1990	ts, composite	-	-	-	-	-	-	-	x?	-	-	-
		PA, 1990	ts, composite	-	-	-	-	-	-	-	-	-	-	x
		W, 1985	composite, ?method	-	-	-	-	-	-	-	-	-	-	x
Carriacou	Cherry Hill Basalt	JBS, 1990	90-46a, site 14	-	x?	-	x?	-	-	-	-	-	-	-
			90-46b, ts, site 14	-	-	-	-	x?	-	x?	x?	-	-	-
	Bogles Limestone	JBS, 1990	90-8b, ts, site 27	-	-	x	-	-	-	-	-	-	-	-
			90-34, ts, site 28	-	-	-	-	x?	-	-	x?	-	-	-
	Belvedere Formation	JBS, 1990	10-9-7-m2, site 29	-	-	-	-	r	-	-	-	-	-	-
			90-47, site 20	-	-	-	-	-	-	-	-	-	-	-
			90-53, site 24	-	-	-	-	-	-	-	-	-	-	-
		W, 1985	Cari 15	-	-	-	-	-	-	-	-	-	-	-
			Cari 16, 17 (vic. Dover, site uncertain)	-	-	-	-	-	-	-	-	-	-	-
		RJ, 1972	10815, 10816	-	-	-	-	-	-	-	-	-	-	-
	Anse La Roche Formation	JBS, 1990	90-28, site 4	-	-	-	-	-	-	-	-	x	-	-
		W, 1985	composite (vic. Gun Point)	-	-	-	-	-	x	-	-	-	-	-
		RJ, 1972	10764, ts	-	-	x	-	-	-	-	-	-	x	-
	Belmont Formation B	RJ, 1972	10740	-	-	-	-	-	-	-	-	-	-	-
			10745	-	-	-	-	-	-	-	-	-	-	-
		W, 1985	Cari 19	-	-	-	-	-	-	-	-	-	-	-
	Kendeace Formation	RJ, 1972	10788, 10790	-	-	-	-	-	-	-	-	-	-	-
			10779	-	-	-	-	-	-	-	-	-	-	-
			10817	-	-	-	-	-	-	-	-	-	-	-
			10699	-	-	-	-	-	-	-	-	-	-	-
			10737	-	-	-	-	-	-	-	-	-	-	-
	Carriacou Formation type section, basal echinoidal limestone	RJ, 1972	10754, 10777, 10788, 10787, all composite samples	-	-	-	-	-	-	-	-	-	-	-
	unit above algal limestone	RJ, 1972	no sample numbers	-	-	-	-	-	-	-	-	-	-	-
Canouan	Canouan Formation	W, 1985	composite, approx. 10 m section	-	-	-	-	-	-	-	-	-	-	-
Mustique	Marls at Black Sand Bay and Gallicaux Bay	W, 1985	Composite of 8 samples	-	-	-	-	-	-	-	-	-	-	-
Union	Sandstone-chert unit	JBS, 1989	ts, composite	-	-	-	-	-	-	-	-	-	-	-
Grenada	Tufton Hall Formation	S, 1985a	composite, at type section	-	-	-	-	x	-	-	-	-	-	-

TABLE 2. PLANKTIC FORAMINIFERS IDENTIFIED IN OLDER ROCKS OF THE GRENADINES AND GRENADA

Island	Unit	Observer, Ref.*	Sample	T. bolivariana	T. frontosa	T. boweri-possango-nensis gr.	T. siakensis	T. opima opima	T. cf opima	T. opima nana	T. mayeri	T. obesa	T. cocoaensis	T. increbescens
Mayreau	Anse Bandeau Formation	JBS, 1990	ts, composite	-	-	-	-	-	-	-	-	-	-	-
		PA, 1990	ts, composite	-	x	-	-	-	-	-	-	-	-	-
		W, 1985	ts, composite	-	-	x	-	-	-	-	-	-	-	-
Jamesby	Marly bed	W, 1985	composite	x	-	-	-	-	-	-	-	-	-	-
Baradel	Chert unit	JBS, 1990	ts, composite	-	-	-	-	-	-	-	-	-	-	-
		PA, 1990	ts, composite	-	-	-	-	-	-	-	-	-	-	-
		W, 1985	composite, ?method	-	-	-	-	-	-	-	-	-	-	-
Carriacou	Cherry Hill Basalt	JBS, 1990	90-46a, site 14	-	-	-	-	-	-	-	-	-	-	-
			90-46b, ts, site 14	-	-	-	-	-	-	-	-	-	-	-
	Bogles Limestone	JBS, 1990	90-8b, ts, site 27	-	-	-	-	-	-	-	-	-	-	-
			90-34, ts, site 28	-	-	-	-	-	-	-	-	-	-	-
	Belvedere Formation	JBS, 1990	10-9-7-m2, site 29	-	-	-	-	-	-	-	-	-	-	-
			90-47, site 20	-	-	-	-	-	-	-	-	-	-	-
			90-53, site 24	-	-	-	-	-	-	-	-	-	-	-
		W, 1985	Cari 15	-	-	-	-	-	x	x	x	x	-	-
			Cari 16, 17 (vic. Dover, site uncertain)	-	-	-	-	-	-	-	-	-	-	-
		RJ, 1972	10815, 10816	-	-	-	-	x	-	-	-	-	-	-
	Anse La Roche Formation	JBS, 1990	90-28, site 4	-	-	-	-	-	-	-	-	-	x	-
		W, 1985	composite, (vic. Gun Point)	-	-	-	-	-	-	-	-	-	x	x
		RJ, 1972	10764, ts	-	-	-	-	-	-	-	-	-	-	-
	Belmont Formation B	RJ, 1972	10740	-	-	-	-	-	-	-	-	-	-	-
			10745	-	-	-	x	-	-	x	-	-	-	-
		W, 1985	Cari 19	-	-	-	-	-	-	-	-	-	-	-
	Kendeace Formation	RJ, 1972	10788, 10790	-	-	-	x	-	-	-	-	-	-	-
			10779	-	-	-	x	-	-	-	-	-	-	-
			10817	-	-	-	-	-	-	-	-	-	-	-
			10699	-	-	-	-	-	-	-	-	-	-	-
			10737	-	-	-	-	-	-	-	-	-	-	-
	Carriacou Formation type section, basal echinoidal limestone	RJ, 1972	10754, 10777, 10778, 10787, all composite samples	-	-	-	-	-	-	-	-	-	-	-
	unit above algal limestone	RJ, 1972	no sample numbers	-	-	-	-	-	-	-	-	-	-	-
Canouan	Canouan Formation	W, 1985	composite, approx. 10-m section	-	-	-	-	-	-	-	x	-	-	-
Mustique	Marls at Black Sand Bay and Gallicaux Bay	W, 1985	composite of 8 samples	-	-	-	-	-	x	x	-	-	-	-
Union	Sandstone-chert unit	JBS, 1990	ts, composite	-	-	-	-	-	-	-	-	-	-	-
Grenada	Tufton Hall Formation	S, 1985a	composite, at type section	-	-	-	-	-	-	x	-	-	-	-

TABLE 2. PLANKTIC FORAMINIFERS IDENTIFIED IN OLDER ROCKS OF THE GRENADINES AND GRENADA (continued)

Island	Unit	Observer, Ref.*	Sample	T. archeom-enardi	T. peripheronda	Catapsydrax sp.	C. dissimilis	C. stainforthi	Casigerinella chipolensis	Globigerina Venezuelana	G. ciperoensis gr.	G. tripartita	G. praebulloides	G. angulisu-turalis
Mayreau	Anse Bandeau Formation	JBS, 1990	ts, composite	-	-	-	-	-	-	-	-	-	-	-
		PA, 1990	ts, composite	-	-	-	-	-	-	-	-	-	-	-
		W, 1985	ts, composite	-	-	-	-	-	-	-	-	-	-	-
Jamesby	Marly bed	W, 1985	composite	-	-	-	-	-	-	-	-	-	-	-
Baradel	Chert unit	JBS, 1990	ts, composite	-	-	-	-	-	-	-	-	-	-	-
		PA, 1990	ts, composite	-	-	-	-	-	-	-	-	-	-	-
		W, 1985	composite, ?method	-	-	-	-	-	-	-	-	-	-	-
Carriacou	Cherry Hill Basalt	JBS, 1990	90-46a, site 14	-	-	-	-	-	-	-	-	-	-	-
			90-46b, ts, site 14	-	-	-	-	-	-	-	-	-	-	-
	Bogles Limestone	JBS, 1990	90-8b, ts, site 27	-	-	-	-	-	-	-	-	-	-	-
			90-34, ts, site 28	-	-	-	-	-	-	-	-	-	-	-
	Belvedere Formation	JBS, 1990	10-9-7-m2, site 29	-	-	-	x	-	-	-	x	-	-	-
			90-47, site 20	-	-	-	x	-	-	-	-	x	-	-
			90-53, site 24	-	-	-	-	-	-	-	-	x	-	-
		W, 1985	Cari 15	-	-	-	x	-	-	-	-	x	x	x	-
			Cari 16, 17 (vic. Dover, site uncertain)	-	-	-	-	-	-	-	-	-	-	x
		RJ, 1972	10815, 10816	-	-	-	-	-	-	-	-	-	x	-
	Anse La Roche Formation	JBS, 1990	90-28, site 4	-	-	-	-	-	-	-	-	-	-	-
		W, 1985	composite, (vic. Gun Point)	-	-	-	-	-	-	-	-	x	-	-
		RJ, 1972	10764, ts	-	-	-	-	-	-	-	-	-	-	-
	Belmont Formation B	RJ, 1972	10740	-	-	x	-	-	x	-	-	-	x	-
			10745	-	-	-	x	-	-	-	-	-	-	-
		W, 1985	Cari 19	-	-	-	-	-	-	-	-	-	-	-
	Kendeace Formation	RJ, 1972	10788, 10790	-	x	-	-	-	-	-	-	-	-	-
			10779	-	-	-	-	-	-	-	-	-	-	-
			10817	-	-	-	-	-	-	-	-	-	-	-
			10699	x	x	-	-	-	-	-	-	-	-	-
			10737	-	-	-	-	-	-	-	-	-	-	-
	Carriacou Formation, type section, basal echinoidal limestone	RJ, 1972	10754, 10777, 10778, 10787, all composite samples	-	-	-	-	-	-	x	-	-	-	-
	unit above algal limestone	RJ, 1972	no sample numbers	-	x	-	-	-	-	x	-	-	-	-
Canouan	Canouan	W, 1985	composite, approx. 10-m section	x	x	-	-	-	-	-	x	-	-	-
Mustique	Marls at Black Sand Bay and Gallicaux Bay	W, 1985	composite of 8 samples	-	-	-	-	-	-	-	x	x	-	-
Union	Sandstone-chert unit	JBS, 1990	ts, composite	-	-	-	-	-	-	-	-	-	-	-
Grenada	Tufton Hall Formation	S, 1985a	composite, at type section	-	-	-	-	-	-	-	-	-	-	-

TABLE 2. PLANKTIC FORAMINIFERS IDENTIFIED IN OLDER ROCKS OF THE GRENADINES AND GRENADA (continued)

Island	Unit	Observer, Ref.*	Sample	G. ampliampertura	G. gortanii	G. praesepsis	G. foliata	G. angustium-bilicata	Globigerinoides senni	G. altiaperturus	G. siacanus	G. cf siacanus	G. trilobus	G. cf trilobus	G. quadrilobus	G. subquadratus
Mayreau	Anse Bandeau Formation	JBS, 1990	ts, composite	-	-	-	-	-	-	-	-	-	-	-	-	-
		PA, 1990	ts, composite	-	-	-	-	-	-	-	-	-	-	-	-	-
		W, 1985	ts, composite	-	-	-	-	-	-	-	-	-	-	-	-	-
Jamesby	Marly bed	W, 1985	composite	-	-	-	-	-	-	-	-	-	-	-	-	-
Baradel	Chert unit	JBS, 1990	ts, composite	-	-	-	-	-	-	-	-	-	-	-	-	-
		PA, 1990	ts, composite	-	-	-	-	-	-	-	-	-	-	-	-	-
		W, 1985	composite, ?method	-	-	-	-	-	-	-	-	-	-	-	-	-
Carriacou	Cherry Hill Basalt	JBS, 1990	90-46a, site 14	-	-	-	-	-	-	-	-	-	-	-	-	-
			90-46b, ts, site 14	-	-	-	-	-	-	x?	-	-	-	-	-	-
	Bogles Limestone	JBS, 1990	90-8b, ts, site 27	-	-	-	-	-	-	-	-	-	-	-	-	-
			90-34, ts, site 28	-	-	-	-	-	-	-	-	-	-	-	-	-
	Belvedere Formation	JBS, 1990	10-9-7-m2, site 29	-	-	-	-	-	-	-	-	-	-	-	-	-
			90-47, site 20	-	-	-	-	-	-	-	-	-	-	-	-	-
			90-53, site 24	-	-	-	-	-	-	-	-	-	-	-	-	-
		W, 1985	Cari 15	-	-	-	-	-	-	-	-	-	-	-	-	-
			Cari 16, 17 (vic. Dover, site uncertain)	-	-	-	-	-	-	-	-	-	-	-	-	-
		RJ, 1972	10815, 10816	-	-	-	-	x	-	-	-	-	-	-	-	-
	Anse La Roche Formation	JBS, 1990	90-28, site 4	-	-	-	-	-	-	-	-	-	-	-	-	-
		W, 1985	composite, (vic. Gun Point)	x	-	-	-	-	-	-	-	-	-	-	-	-
		RJ, 1972	10764, ts	-	-	-	-	-	-	-	-	-	-	-	-	-
	Belmont Formation B	RJ, 1972	10740	-	-	-	x	-	-	-	-	-	-	-	-	-
			10745	-	-	-	-	-	-	-	-	-	-	x	-	-
		W, 1985	Cari 19	-	-	-	-	-	-	-	x	-	-	-	-	-
	Kendeace Formation	RJ, 1972	10788, 10790	-	-	-	-	-	-	-	-	-	x	-	-	-
			10779	-	-	-	-	-	-	-	-	-	x	-	-	x
			10817	-	-	-	x	-	-	-	-	x	x	-	-	-
			10699	-	-	-	-	-	-	-	-	-	-	-	x	-
			10737	-	-	-	-	-	-	-	x	-	x	-	-	-
	Carriacou Formation, type section, basal echinoidal limestone	RJ, 1972	19754, 10777, 10778, 10787, all composite samples	-	-	-	-	-	-	-	-	x	x	-	-	-
	unit above algal limestone	RJ, 1972	no sample numbers	-	-	-	-	-	-	-	x	-	-	-	-	-
Canouan	Canouan	W, 1985	composite, approx. 10-m section	-	-	-	-	-	-	-	x	-	-	-	-	-
Mustique	Marls at Black Sand Bay and Gallicaux Bay	W, 1985	composite of 8 samples	-	x	x	-	-	-	-	-	-	-	-	-	-
Union	Sandstone-chert unit	JBS, 1990	ts, composite	-	-	-	-	-	-	-	-	-	-	-	-	-
Grenada	Tufton Hall Formation	S, 1985a	composite, at type section	-	-	-	-	-	-	-	-	-	-	-	-	-

TABLE 2. PLANKTIC FORAMINIFERS IDENTIFIED IN OLDER ROCKS OF THE GRENADINES AND GRENADA (continued)

Island	Unit	Observer, Ref.*	Sample	G. immaturus	G. ruber	G. diminutus	Globoquadrina altispira	G. dehiscens	Globigerinatella insueta	Praeorbulina glomerosa	Catapsydrax unicavus	Turborotalia centralis
Mayreau	Anse Bandeau Formation	JBS, 1990 PA, 1990 W, 1985	ts, composite ts, composite ts, composite	- - -	- - -	- - -	- - -	- - -	- - -	- - -	- - -	- - -
Jamesby	Marly bed	W, 1985	composite	-	-	-	-	-	-	-	-	-
Baradel	Chert unit	JBS, 1990 PA, 1990 W, 1985	ts, composite ts, composite composite, ?method	- - -	- - -	- - -	- - -	- - -	- - -	- - -	- - -	- - -
Carriacou	Cherry Hill Basalt	JBS, 1990	90-46a, site 14 90-46b, ts, site 14	- -	- -	- -	- -	- -	- -	- -	- -	- -
	Bogles Limestone	JBS, 1990	90-8b, ts, site 27 90-34, ts, site 28	- -	- -	- -	- -	- -	- -	- -	- -	- -
	Belvedere Formation	JBS, 1990 W, 1985 RJ, 1972	10-9-7-m2, site 29 90-47, site 20 90-53, site 24 Cari 15 Cari 16, 17 (vic. Dover, site uncertain) 10815, 10816	- - - - - -	- - - - - -	- - - - - -	- - - - - -	- - - - - -	- - - - - -	- - - - - -	- - - - - -	- - - - - -
	Anse La Roche Formation	JBS, 1990 W, 1985 RJ, 1972	90-28, site 4 composite, (vic. Gun Point) 10764, ts	- - -	- - -	- - -	- - -	- - -	- - -	- - -	- - -	- - -
	Belmont Formation B	RJ, 1972 W, 1985	10740 10745 Cari 19	- x -	- - -	- - -	- x x	- x -	- - x	- - -	- - -	- - -
	Kendeace Formation	RJ, 1972	10788, 10790 10779 10817 10699 10737	- - - - -	- - - - -	- - - - -	x - - x -	- x x x -	x - - - -	- - - - -	- - - - -	- - - - -
	Carriacou Formation, type section, basal echinoidal limestone	RJ, 1972	10754, 10777, 10778, 10787, all composite samples	-	-	-	x	-	x	-	-	-
	unit above algal limestone	RJ, 1972	no sample numbers	-	-	-	-	-	-	x	-	-
Canouan	Canouan	W, 1985	composite, approx. 10-m section	-	x	x	x	x	-	x	-	-
Mustique	Marls at Black Sand Bay and Gallicaux Bay	W, 1985	composite of 8 samples	-	-	-	-	-	-	-	-	-
Union	Sandstone-chert unit	JBS, 1990	ts, composite	-	-	-	-	-	-	-	-	-
Grenada	Tufton Hall Formation	S, 1985a	composite, at type section	-	-	-	-	-	-	-	x	x

TABLE 2. PLANKTIC FORAMINIFERS IDENTIFIED IN OLDER ROCKS OF THE GRENADINES AND GRENADA (continued)

Island	Unit	Observer, Ref.*	Sample	T. cunialensis	Hantkenina suprasuturalis	Cribohantkenina danvillensis	Foraminifer Zonal Range
Mayreau	Anse Bandeau Formation	JBS, 1990	ts, composite	-	-	-	upper P10-P11
		PA, 1990	ts, composite	-	-	-	upper P10-P11
		W, 1985	ts, composite	-	-	-	upper P10-P11
Jamesby	Marly bed	W, 1985	composite	-	-	-	P12
Baradel	Chert unit	JBS, 1990	ts, composite	-	-	-	P11
		PA, 1990	ts, composite	-	-	-	P11
		W, 1985	composite, ?method	-	-	-	P11
Carriacou	Cherry Hill Basalt	JBS, 1990	90-46a, site 14	-	-	-	P10-12?
			90-46b, ts, site 14	-	-	-	P12-14?
	Bogles Limestone	JBS, 1990	90-8b, ts, site 27	-	-	-	P10-14
			90-34, ts, site 28	-	-	-	P12-14?
	Belvedere Formation	JBS, 1990	10-9-7-m2, site 29	-	-	-	P13-17
			90-47, site 20	-	-	-	P13-N5
			90-53, site 24	-	-	-	P18-N5
		W, 1985	Cari 15	-	-	-	top P21
			Cari 16, 17 (vic. Dover, site uncertain	-	-	-	P20-22
		RJ, 1972	10815, 10816	-	-	-	P20-21
	Anse La Roche	JBS, 1990	90-28, site 4	-	-	-	P15-17
		W, 1985	composite, (vic. Gun Point)	-	-	-	upper P16-P17
		RJ, 1972	10764, ts	-	-	-	P10-P17
	Belmont Formation B	RJ, 1972	10740	-	-	-	P19-N7
			10745	-	-	-	N5-6
		W, 1985	Cari 19	-	-	-	N5-6
	Kendeace Formation	RJ, 1972	10788, 10790	-	-	-	mid N6-N8
			10779	-	-	-	P21-N15
			10817	-	-	-	P21-N17
			10699	-	-	-	N6-N10
			10737	-	-	-	P21-N21
	Carriacou Formation, type section, basal echinoidal limestone	RJ, 1972	10754, 10777, 10778, 10787, all composite samples	-	-	-	N7-8
	unit above algal limestone	RJ, 1972	no sample numbers	-	-	-	N8
Canouan	Canouan	W, 1985	composite, approx. 10-m section	-	-	-	N5
Mustique	Marls at Black Sand Bay and Gallicaux Bay	W, 1985	composite of 8 samples	-	-	-	P19-21
Union	Sandstone-chert unit	JBS, 1990	ts, composite	-	-	-	P10-15
Grenada	Tufton Hall Formation	S, 1985a	composite, at type section	x	x	x	P17

*JBS, 1990 = John B. Saunders, this paper; PA, 1990 = Patrick Andreieff, written communication, 1990; W, 1985 = Westercamp and others, 1985; RJ, 1972 = Robinson and Jung, 1972; S,1985a = Saunders and others, 1985a; ts = thin section identification, all others are from disaggregated samples; x = present; r = rich; ? = uncertain; Carriacou samples sites on Figures 7, 10, and 15; foram zonal nomenclature from Blow, 1969.

packets evidently took up distributed slip during disruption of the folded layers.

The motions in folding and disruption were almost certainly related. Where intermediate limbs of folds are preserved (Fig. 6H), vergence is northerly regardless of whether axial planes dip north or south. The prevalence of upright beds in folds and in phacoids suggests northerly vergence was general during folding. Offsets and striations on faults in the disrupted zone indicate updip transport (Fig. 6L, O) and that the earlier faults are mainly thrusts. Thus, zones of disrupted, folded beds underwent early northward overriding and tectonic thickening. The short length of overturned limbs suggests that faulting occurred at small fold amplitude or that some folds could have formed or amplified during faulting by local sticking or ramping of the hangingwall.

The later north-dipping normal faults are more widely spaced and less branched than the earlier faults (Fig. 6C). Their slip is downdip, and throw is northerly. Thus, the two fault sets were not conjugates in a coaxial deformation. The normal faulting may have caused rotation to more southerly dips of the earlier structures in the zones of disruption.

Tectonic fabrics. Orientation diagrams (Fig. 6) indicate reasonably systematic orientation ranges of structural elements in the Eocene turbidites. Fold axes plunge shallowly east or west with azimuthal range <50°. Axial planes and axial plane cleavage are mainly south dipping but are partly girdled about the mean axial direction. Fault planes of both generations occupy much the same partial girdle as bedding, as does net slip. Thus, the diagrams indicate a monoclinic fabric: a constant east-west–trending axis of folding and rotation of surfaces in the plane of the girdle (north-south). The orientation of the total fabric varies a little with position in the Eocene turbidites. For example, the fabric orientation varies from eastern Levera (Fig. 6F, G) to western Levera (Fig. 6K, L), a distance of 0.5 to 1 km. The orientation difference can be explained by a small rotation about a moderately north-plunging axis.

Interpretation of structures. The three principal structural associations in the Tufton Hall Formation are in fact probably gradational because bedding-parallel faults and fault-related folds occur in all the associations and because homoclines may be the upright limbs of major recumbent folds. Disruption exists to some degree in all associations but is concentrated in successions where minor folds are abundant. We interpret the disruption to have occurred during the early development of local trains of asymmetric minor folds. The branching thrusts that caused disruption may be related to the bedding-subparallel shears that affected successions withour minor folds.

Deformation of the Eocene turbidites evolved in two main phases. First was the generation of north-verging folds and thrusts and the disruption of some folded rocks. Such motions evidently included north-south contraction together with northward over-riding and variable internal rotation within fault packets. Second was north-south extension that caused low- and high-angle normal faulting. Two other poorly resolved kinematic elements are (1) possible bedding-parallel cleavage which may suggest that flattening due to tectonics or compaction preceded what is called first-phase structures, and (2) the small rotation of total fabric between study areas; the latter element must represent the youngest deformational event.

The environment of deformation of the Eocene turbidites was nonmetamorphic and probably at shallow depths if not surficial. This is indicated by fold styles, the low degree of foliation development, and the importance of faulting in all stages and all structural associations. There are no evident strains of sand-sized and coarser particles; displacements were thus taken up by slip on layer and grain boundaries in sandy rocks and by modest strains in muddy rocks. There is no evidence, however, that the Eocene turbidites were unlithified during deformation or that deformation occurred during sedimentation.

The heterogeneity of deformation of the Tufton Hall Formation can be explained in at least two ways. First, if the initial deformation was principally by thrust imbrication rather than by folding, the zonation of structures may reflect heterogeneity in rheologic properties within the strata (initial water content, permeability, bedding-plane adhesions, larger stiffness by cementation). Alternatively, the strata may have initially deformed in a train of north-verging folds with a spectrum of wavelengths that accorded to rheologic variations in the section; many folds then faulted in their intermediate limb, leading to three different structural associations.

The base of the Tufton Hall Formation is nowhere exposed. It is unknown whether this unit is parautochthonous or allochthonous.

Other older rocks. Aside from the Tufton Hall Formation, rocks older than the Neogene volcanics are known only in two small outcrop areas east of St. George's and in blocks in tuff near Brizen (Fig. 6A). It is uncertain whether the outcrops east of St. George's are in windows through the Neogene volcanics or are huge blocks within the magmatic cover.

In the vicinity of the communities of Tempe and Mt. Parnassus (Fig. 6A) are outcrops of well-bedded, sediment-gravity flows and a few hemipelagic interbeds. The predominant coarse sandy and pebbly grain flows contain skeletal and volcanigenic particles. Saunders and others (1985a, b) obtained foraminifers from a marl in the Tempe-Parnassus series that yield a zonal age of P19, *Turborotalia opima*, middle Oligocene. The beds are folded in a close macroscopic anticline with subhorizontal axis that trends east-west. The anticline is >100 m wide and can be traced 200 m on axis. If this unit is not in a large rotated block, the fold orientation implies that north-south contraction in Grenada either began after or continued through middle Oligocene time.

The outcrop area of older rocks near Hope Vale (Fig. 6A) has not been investigated by us. Saunders and others (1985a) state it contains limy rocks and has yielded foraminifers that give a permissible age range of middle Oligocene to middle Miocene.

The limestone blocks at Brizen gave larger foraminifers thought to be Oligocene (Saunders and others, 1985a).

Dating of Neogene andesite at Mt. Craven. The Northern Domes center of eruptive rocks (Fig. 6B) was differentiated by Arculus (1976) among five centers that constitute the principal Neogene volcanoes of Grenada. The Northern Domes center includes several domes of andesite and basaltic and andesitic lavas. Stratigraphic evidence for the ages or sequence of the centers is unavailable in exposures, but it is clear that the volcanic rocks are post-Eocene by virtue of their superposition on the Tufton Hall Formation. A sequence of the eruptive centers was erected, however, by K-Ar dates (Briden and others, 1979). Of these, the oldest (21.2 Ma) was from hornblende in the andesite dome at Mt. Craven (Fig. 6) of the Northern Domes center, implying that center is the oldest on Grenada (Arculus, 1976).

We have redated hornblende from the dome at Mt. Craven by the $^{40}Ar/^{39}Ar$ technique, which provides an age-heating step spectrum (Fig. 5B to E; Table 1). We sampled the same outcrop that gave the specimen dated by Briden and others (1979) via directions by R. J. Arculus. The analyses are by K. A. Foland. Hornblende exists as distributed phenocrysts of variably stubby and elongate prismatic forms, mainly well lineated with subvertical plunge, in a microcrystalline felsic groundmass. The hornblende is clearly a magmatic phase, not xenolithic nor in cognate inclusions. It is red-brown and has no alteration by weathering or metamorphism in the thin section examined.

We dated outcrop samples of hornblende andesite from two sites about 200 m apart (91-1 and 91-4, Fig. 6B). Each sample consisted of chips of fresh rock taken over a 10 m^2 surface. Hornblende was separated from each sample and divided into two fractions, each of which was analyzed. The four spectra (Fig. 5B to E) are similar and show the following features: (1) central plateaus that cover 70% or more of the argon release with similar dates, between 1.34 and 1.79 Ma, and very small uncertainty; (2) older argon release at the beginning and final increments, amounting to no more than 10% of total argon; (3) total gas dates between 1.9 and 2.3 Ma; and (4) uniformly low equivalent potassium content, about 0.35 wt. % K.

The similar spectral characteristics among the four analyses of hornblende indicate the small range of dates includes the age of crystallization and nearly synchronous cooling through the blocking temperature (about 550°C for hornblende). The argon released at high and low temperature ends of the spectra is erratic in date versus release temperature, including dates as old as 46 Ma. Such argon released at the low temperature end is almost certainly extraneous, not a product of decay in place. The old argon released at the high temperature end may also be extraneous or conceivably inherited if the hornblende phenocrysts are intratelluric.

To conclude, the hornblende andesite at Mt. Craven was probably emplaced and cooled between about 1.3 and 1.8 Ma. The Northern Domes center is of post–mid-Pliocene age and is not necessarily the oldest volcanic center of Grenada. The K-Ar date of 21.2 Ma of Briden and others (1979) is an invalid age and no longer gives the minimum age for the onset of Neogene volcanism in the SLAAP.

Carriacou

Introduction. Carriacou is widely underlain by Paleogene and Neogene strata that provide the most complete pre-late Miocene record of the SLAAP (Fig. 7). The strata were assigned to formations and dated in part (Fig. 8a) by Martin-Kaye (1958, 1969) and Robinson and Jung (1972) and mapped by Jackson (1970) and by us. Neogene magmatic rocks, mainly or wholly intrusive, are younger than the stratiform rocks. From ten sampling sites in such magmatic rocks, the range of ages (K-Ar, whole rock) is 2.7 to 11.2 Ma (Fig. 9; Briden and others, 1979). It is not certain, however, that the oldest Neogene magmatic rocks are dated.

From new field work and dating, we present a substantially different organization of rock units from that of prior work (Fig. 8a), depicted as a tectonostratigraphy in Figure 8b. Figure 9 gives known age ranges of units named in Figure 8b. The part of northern Carriacou in which our studies are concentrated is shown in map and sections at large scale in Figure 10. Tables 2 and 3 list species of foraminifers and nannofossils, respectively, from dated specimens.

Major findings from Carriacou are as follows. (1) Paleogene strata of different facies are thrust together, probably with large displacement. (2) The allochthon of Paleogene strata includes the newly discovered Cherry Hill Basalt that is among the oldest dated rocks of the SLAAP and almost certainly correlative with the Mayreau Basalt. (3) All rocks younger than the Cherry Hill Basalt (middle Eocene) and older than middle Miocene are sub-wavebase, perhaps deep marine, sedimentary rocks; none is magmatic except for minor basalt near the top (upper Oligocene) of the Belvedere Formation (Fig. 8b).

Stratigraphic changes. The reorganization of rock units of Carriacou introduced here is based on lithostratigraphy, accompanied by dating by nannoflora and foraminifers. The changes are as follows (Fig. 8). (1) The Belvedere Formation, newly defined in this paper, is a distinctive middle Eocene to upper Oligocene or younger succession of marl and volcanigenic turbidite. It includes the here-abandoned Windward Limestone and Tuff of Robinson and Jung (1972). The Belvedere lies unconformably below the Belmont Formation (lower Miocene, possibly upper Oligocene) and with probable depositional contact above new middle Eocene units, the Bogles Limestone and Cherry Hill Basalt (Fig. 8b). The Cherry Hill, Bogles, and Belvedere units compose the Bogles allochthon (Figs. 7, 8b, 10).

The Anse La Roche Formation of Robinson and Jung (1972) is now recognized to occupy a much greater area in western coastal Carriacou and to be the autochthon of the Bogles thrust (Figs. 7, 10). The Anse La Roche was formerly known to be upper Eocene but is now thought to extend into the early Oligocene. The Anse La Roche and Belvedere Formations are synchronous in part (Fig. 9). They are juxtaposed by the Bogles thrust, the displacement along which is unknown but probably more than a few kilometers.

The Kendeace Formation is here defined and elevated from a member of the Belmont Formation, as defined by Robinson and Jung (1972). Reasons for this change are that the Kendeace is widely mappable between the Belmont and Carriacou Formations, has distinctive rocks and is locally unconformable on the Belmont.

We affirm that at least a part of the Belmont, and the Kendeace, Carriacou, and Grand Bay Formations compose a struc-

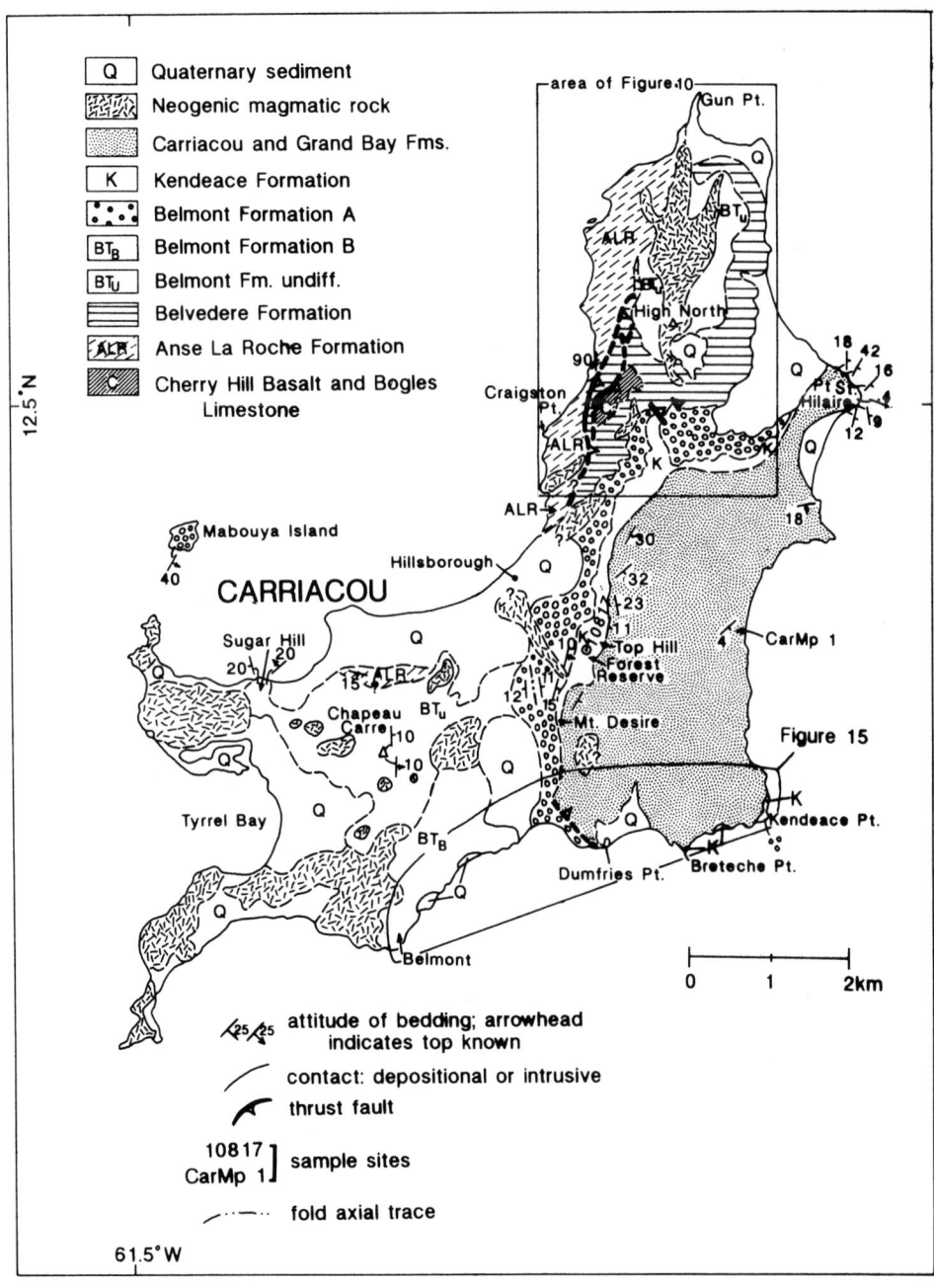

Figure 7. Geologic map of Carriacou; by P. L. Smith and R. Speed, partly after Jackson (1970).

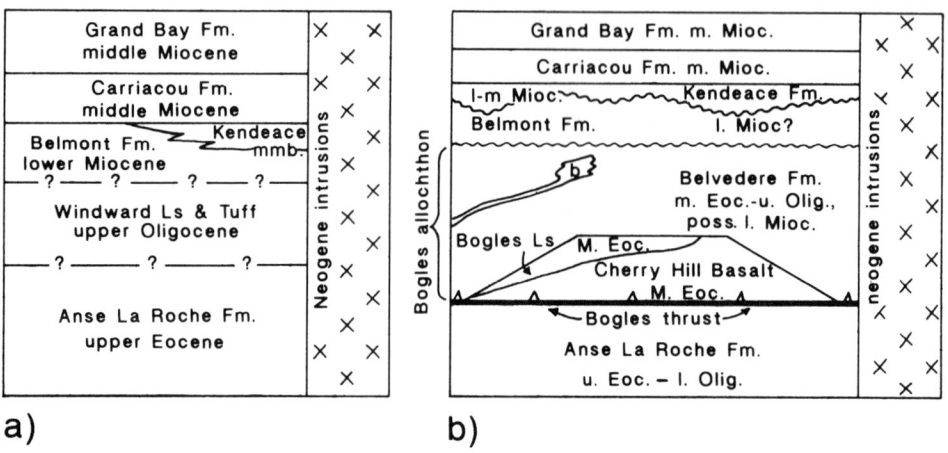

Figure 8. a) Stratigraphic succession of Carriacou from Robinson and Jung (1972). b) Tectonostratigraphy of Carriacou from this study; b is basalt.

turally continuous Miocene succession. An unconformity exists between the Belmont and Kendeace Formations. The Kendeace, Carriacou, and Grand Bay may be a continuous depositional succession, but dating is not well enough resolved to indicate whether small lacunas exist between or within these formations.

Cherry Hill Basalt. This new formation contains pillow basalt and pillow breccia and minor interstitial sediments. The formation is the stratigraphically lowest unit of the Bogles allochthon. Its depositional base is unexposed and may not exist in the subsurface of Carriacou. Its top is probably conformable with the patchy Bogles Limestone contact and with the Belvedere Formation (Fig. 8b). The attitudes of pillows in roadcuts of the Bogles-Belvedere road (Fig. 10) suggest the Cherry Hill Basalt is moderately folded and upright. Its exposed thickness is hard to assess, but at least a few hundred meters (Fig. 10).

We have identified poorly preserved nannofossils and foraminifers in disaggregates and thin section of two specimens of interstitial limestone from the upper few meters of the Cherry Hill Basalt (Tables 2, 3). Taken together, a zonal age of base NP16, equivalently lower P12 (*M. lehneri* zone), is delineated by the first appearances of *R. umbilica* and *H. alabamensis* and by the last appearances of *S. spiniger* and *H. mexicana*. This implies a middle middle Eocene age, between about 45 and 46 Ma, for the interstitial limestone (Fig. 9).

The outcrop area of the Cherry Hill Basalt (Fig. 10) is in an apparent culmination of a macroscopic anticline with east-dipping axial plane and doubly but shallowly (≤10°) plunging axes to the northeast and southwest. The data on which this structure is based are (1) the form surface of the base of suprajacent formations, taken to be a depositional contact, (2) bedding within the suprajacent formations, and (3) meager data on structures of sediments interstitial to pillows of the Basalt (Fig. 11).

Rocks of the Cherry Hill Basalt as a whole have undergone substantial diagenesis and alteration but no recognized metamorphism at greenschist or higher facies. Basalt in pillows or fragments is unfoliated, but interstitial sediment, both hyaloclastite and carbonate, contains spaced cleavage.

The formation contains zones of whole pillows and of pillow breccia. The arrangement of these, as successions or facies, is unclear. Pillow zones are at least tens of meters thick. Pillows are equant to elongate, as much as 5:1; maximum length is about 2 m. The pillow interstices include cleaved dark chloritic sediment, probably of hyaloclastic protoliths, and more conspicuous red limestone and silicified limestone. The replacement origin of the siliceous interstitial sediment is inferred from existence of foram relics and lamination similar to those of calcareous interstitial sediment.

Zones of pillow breccia contain a framework of angular basalt fragments with partial concentric zonation and rounded margins. Such zones also include fragments of interstitial sediment.

The basalt is microporphyritic; small pyroxene phenocrysts exist in a felted feldspathic groundmass. It is highly vesicular and amygdaloidal. The amygdules are spherical or irregularly shaped and contain carbonate ± chlorite.

Structures recognized within this poorly exposed formation are spaced cleavage, a train of mesoscopic folds in interstitial sediments, and brittle faults and fractures of several sets that cut across pillow and matrix. The cleavage (Fig. 11a) and axial planes of the folds dip shallowly to moderately east suggesting they are kinematically related. We infer the interstitial structures were caused by slip between the apparently rigid pillows and basalt fragments. The slip was probably an accommodation to the folding of the Cherry Hill Basalt.

The Cherry Hill Basalt evidently represents marine basaltic extrusion during middle Eocene time at depths shallow enough for substantial vesiculation and preservation of pelagic carbonate (<3.5 km). The interstitial carbonate is interpreted as contemporaneous seabed cover that was intruded by the horizontally migrating pillow front. The pillow breccias may be from a collapsed front of a pillowed flow. The end of basaltic extrusion in middle Eocene time is recorded by the cover of the Cherry Hill Basalt by the Bogles Limestone, whose similar petrography to that of interstitial limestone in the Basalt implies pelagic conditions continued after cessation of extrusion.

The Cherry Hill Basalt is almost certainly correlative with the Mayreau Basalt, exposed 15 km north, by virtue of lithology and association with internal and covering pink pelagic limestone, although the age of cessation of volcanism of the two units differs by 0.5 to 6 m.y. The exposed Cherry Hill Basalt is too weathered to permit chemical evaluation of the tectonics of its origin by surface sampling. By correlation with the Mayreau Basalt, however, we infer it emerged at a spreading center rather than as an arc magma.

Bogles Limestone. This newly defined unit consists of 10 to 30 m of limestone and chert that cap parts of the Cherry Hill Basalt (Figs. 8, 10). The rocks are products of diagenesis of pelagic carbonate and are similar in lithotype and age to interstitial carbonate rock of the Cherry Hill Basalt.

Bedding within the Bogles Limestone is approximately parallel to the contact with the Cherry Hill Basalt and to bedding in the suprajacent Belvedere Formation at the few places where exposures permit observations (Fig. 10). On this basis, we assume such concordancy did or does exist everywhere and that the Basalt and Limestone are a depositional succession. We are uncertain, however, whether the Bogles Limestone and basal Belvedere Formation, which locally caps the Cherry Hill Basalt, are either a succession with erosionally unconformable contact or facies of different diagenetic maturity. We differentiate the Bogles Limestone to call attention to its existence as a probable correlative of the Anse Bandeau Formation of Mayreau.

Rocks of Bogles Limestone are chiefly thin-bedded red planktic foraminiferal, radiolarian micrite. The micrite beds are microcrystalline and mainly massive, but locally depositional lamination and (or) spaced cleavage are evident. They contain virtually no clastic or volcanigenic particles. The principal variant of red micrite is bedded colorless or light green cherty rock that arose by bleaching and silicification of micrite in the walls of fractures. Such cherts contain relics of the same planktic forams as the micrites. More rarely, the Bogles Limestone includes cemented chalk in association with crystalline micrite; the chalk is softer and more porous than the micrite and may be a less diagenetically mature sample of the unit's protolith.

We recovered nannofossils from three samples and foraminifers from two samples of Bogles Limestone (Tables 1, 3). The foram-bearing samples yielded no nannofossils and vice versa. Using the five zonal ranges discretely, the permissible age range is NP14 to 17 or P10 to 14 (Fig. 9). Assuming the formation's age range is delimited by the combined species for the five samples, however, we obtain base NP16 or P12, middle middle Eocene. The latter rationale seems justified by the thinness of the Bogles Limestone. Further the unit's stratigraphic age is within NP16 according to dating of formations above and below.

The Bogles Limestone is derived from pelagic biogenic sediments that accumulated above the mid-Eocene CCD (3.5 km) without influx of terrigenous or volcanigenic particles. The unit probably records the cessation of volcanism that gave rise to the Cherry Hill Basalt because the Bogles Limestone contains no dikes of rock like that of the Basalt. The age or ages of diagenesis of the Bogles Limestone are uncertain and could be both Paleogene and Neogene.

Belvedere Formation. Introduction. The Belvedere Formation is named for a succession of calcareous pelagic and turbiditic rocks of middle Eocene to late Oligocene and possibly younger age (Figs. 7, 8b, 9). The formation is allochthonous on the Bogles thrust (Fig. 10). It has a probably depositional though locally faulted base on the Cherry Hill Basalt and Bogles Limestone and is unconformable below the Belmont Formation of probable lower Miocene age (Figs. 9, 10). The Belvedere Formation is moderately to mildly deformed, depending on proximity to the Bogles thrust. It is unmetamorphosed and contains no foliation or tectonitic fabrics. The Belvedere includes in northeastern Carriacou the Windward Limestone and Tuff of Robinson and Jung (1972) and elsewhere, strata not previously recognized. We recommend use of the new name and call for suppressionn of Windward to emphasize the much larger areal extent and time range of a probably coherent stratal succession than previously known.

General character. The macroscopic structure of the Belvedere is complex; we introduce it here to provide a geometric context for lithic and stratigraphic information that follows. The fundamental structure is a macroscopic dominant anticline with curved axial trace trending north-northeast in western Carriacou and east-northeast in eastern (Fig. 10). The anticline has a culmination at Bogles village (Fig. 10) where the subjacent formations are exposed and away from which its axes plunge in opposite directions. The western limb of the anticline is subvertical west of the culmination and proximal to the Bogles thrust (sections, Fig. 10). This indicates the fold is nearly overturned to the west, at least where it is north-northeast trending and at basal levels in the allochthon. An easterly dip of the axial surface of the macroscopic anticline in the Bogles area may also be indicated by the east-dipping cleavage in the Cherry Hill Basalt (Fig. 11a). Moreover, the western limb is overthrust by the core of the anticline at its culmination (Fig. 10) as detected by the evident truncation of bedding in the Belvedere Formation by the contact with subjacent units at Bogles village. We infer the thrust that cuts the anticline branches from the Bogles thrust at depth (Fig. 10). The macroscopic anticline is cut by the shallowly dipping unconformity that is the Belvedere-Belmont contact (Fig. 10). The curvature of the anticline's axial trace is inferred to be caused by a post-Belmont superposed macroscopic fold with NS to NNW-SSE axial trace that is just west of Belvedere (Fig. 10).

The Belvedere Formation contains generally well bedded interstratified pelagic and turbiditic volcanigenic rocks. The pelagic beds, nearly all calcareous, are mainly marl and range from chalk to mudstone. The volcanigenic sediments are chiefly sandstone and pebbly sandstone and rarely, conglomerate. Minor con-

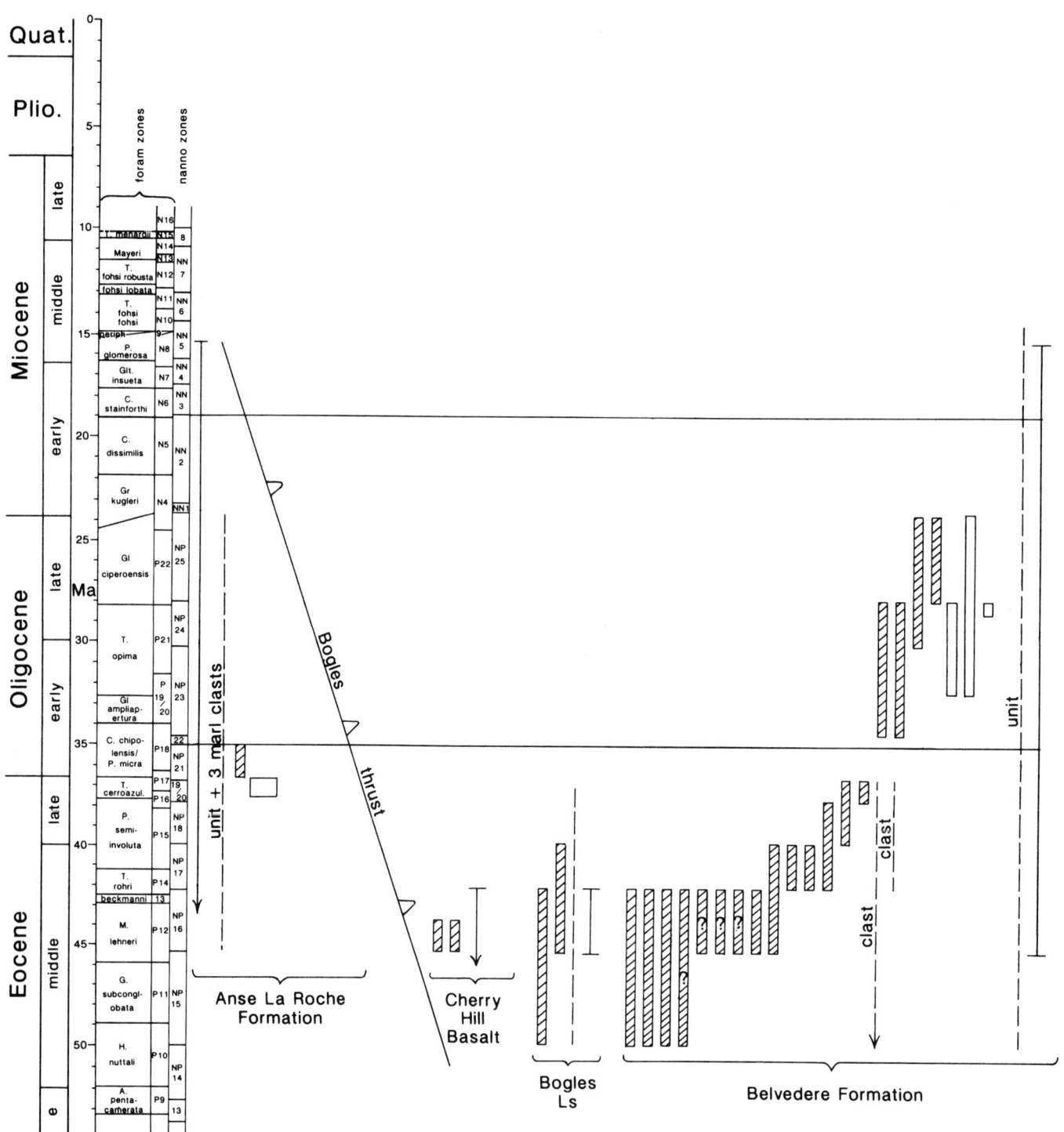

Figure 9 (on this and facing page). Age data for Carriacou; age nomenclature defined in text; foraminifera zones from Blow (1969); nannofossil zones from Martini (1971); correlations among numerical and stratigraphic time scales and zonal ages from Berggren and others (1985).

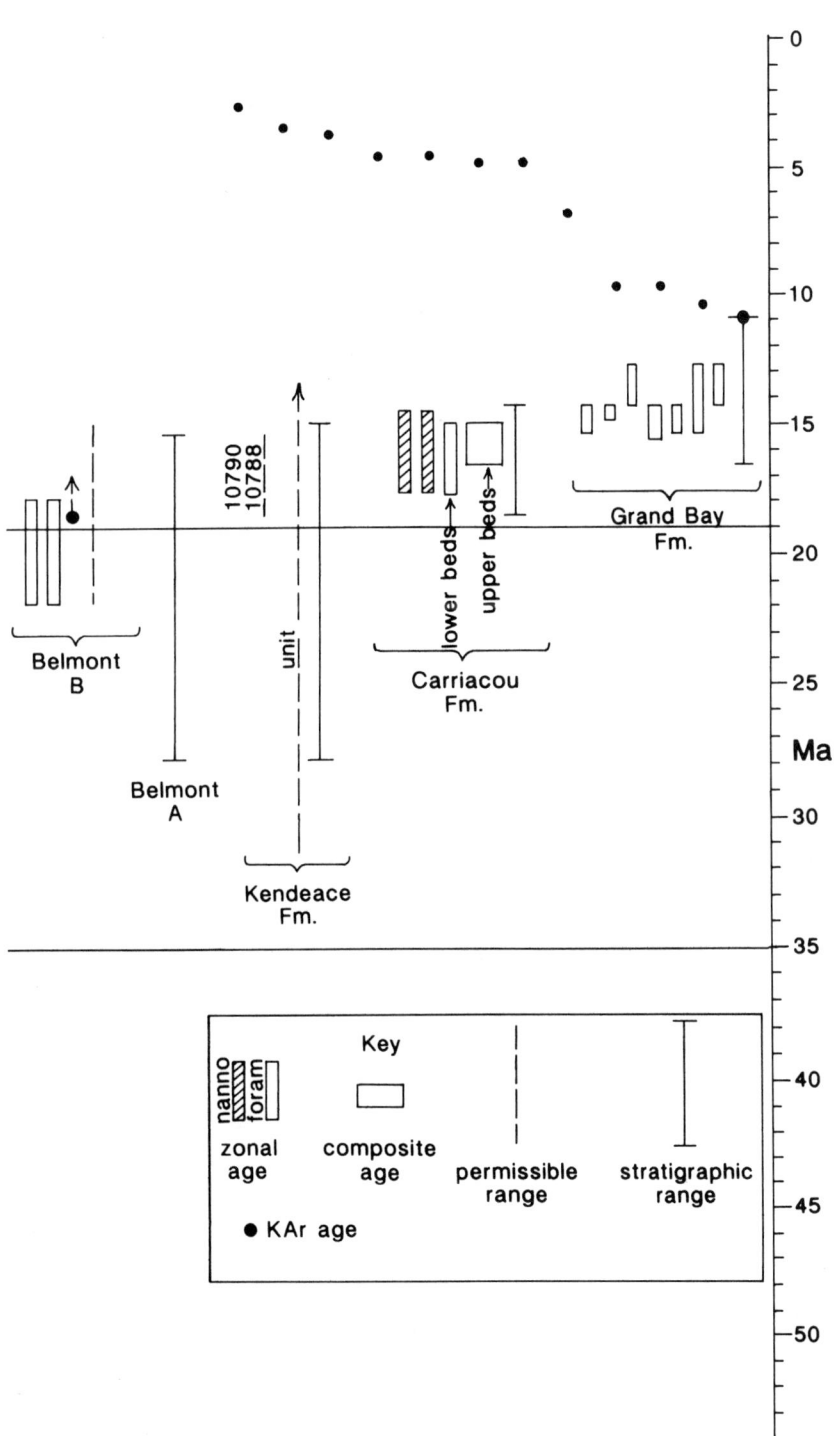

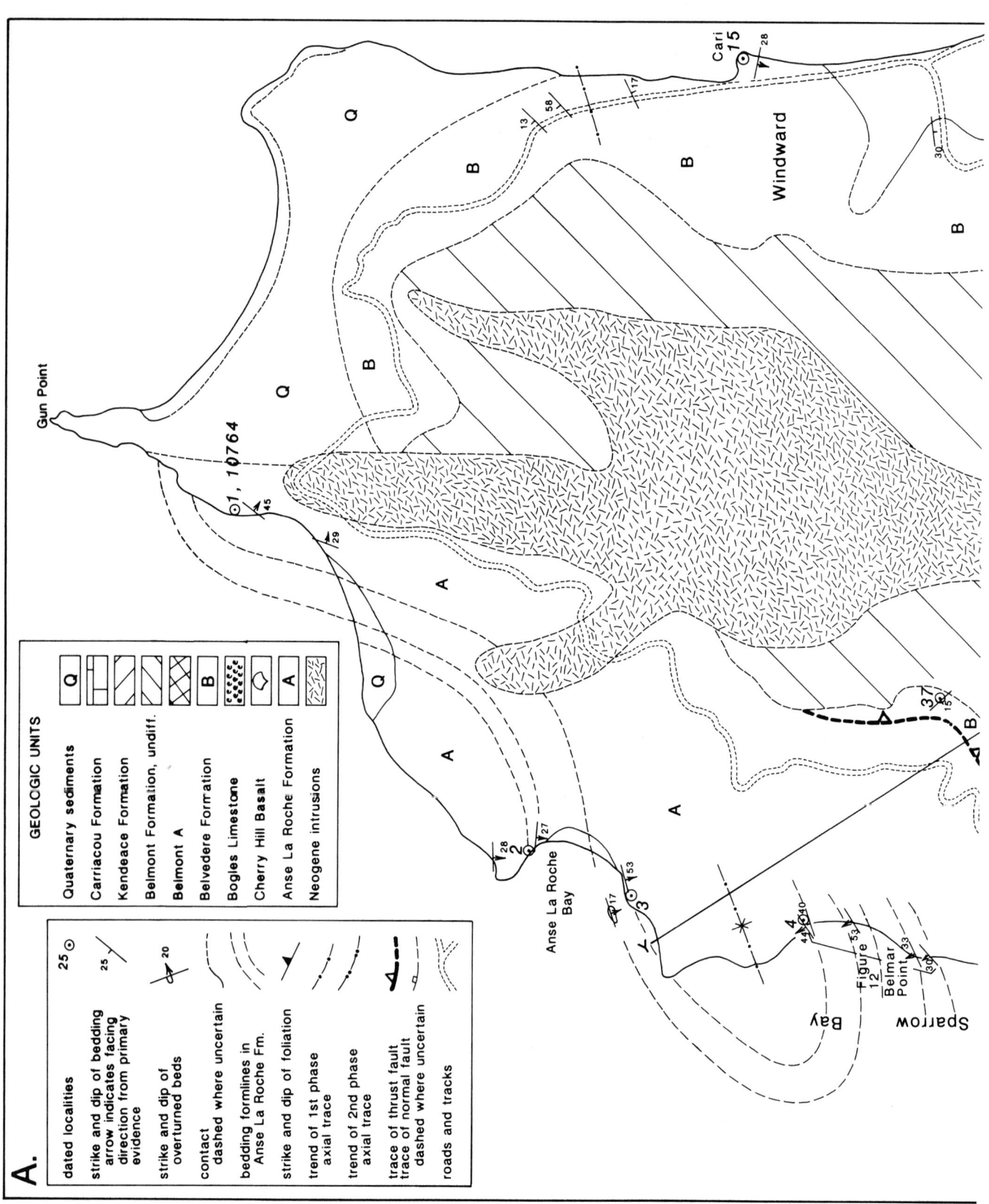

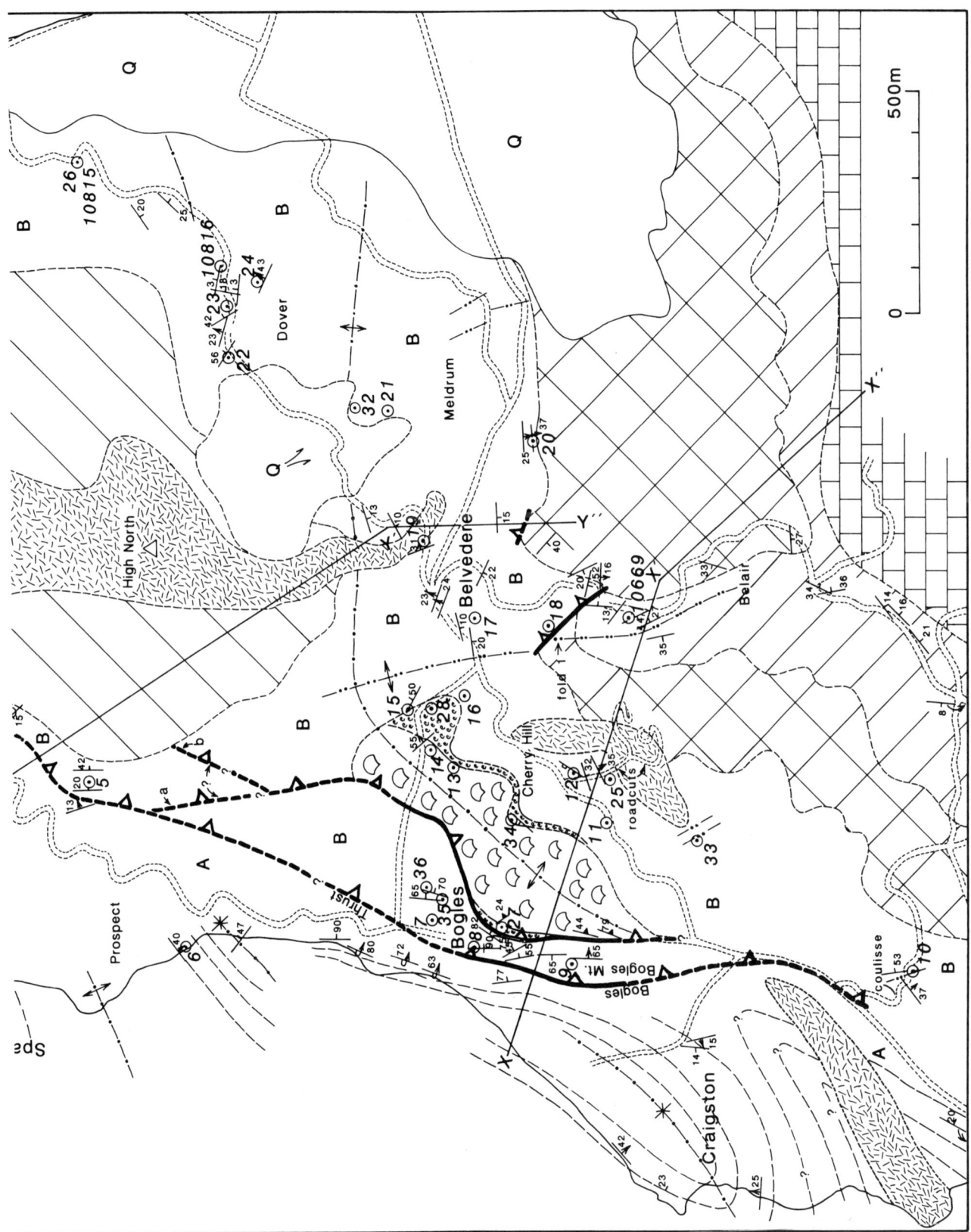

Figure 10. A, Geologic map and B (on following page), sections of northern Carriacou; located on Figure 7; by P. L. Smith and R. Speed, partly after Jackson (1970).

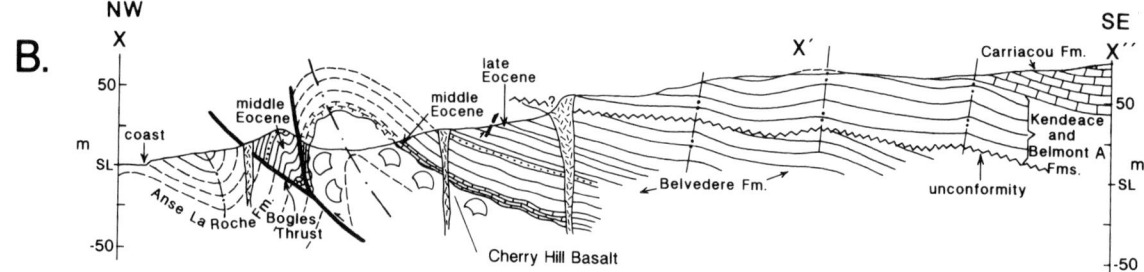
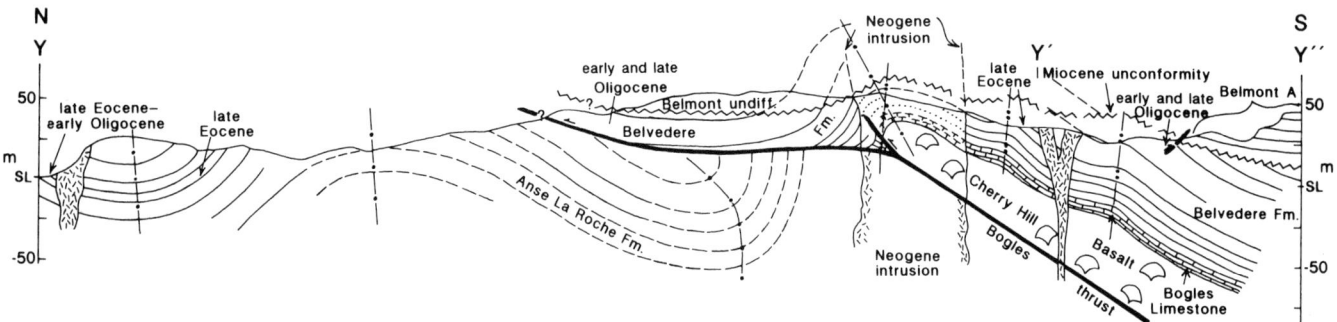

stituents are foram microconglomerate and, near the top of the formation, basalt.

Below, we divide more specific discussion of lithic content, ages, and structure among five areas (Fig. 11): (1) Bogles village, (2) Belvedere southwest, (3) Dover, (4) Windward, and (5) west of High North.

Bogles Village. This area includes the Belvedere Formation in the western limb of the macroscopic anticline where the beds are subvertical and probably repeated in tight minor folds (Fig. 10). No resolvable stratigraphic sequence exists. Rocks include: soft colorless chalk and marl, soft tan massive mudstone with minor calcite, crystalline thin-bedded siliceous micrite, and volcanigenic sandstone and pebbly sandstone. The crystalline micrite is laminated and contains replaced microfossils and quartz veins of submillimeter thickness. The micrite is well layered and almost certainly a product of locally advanced diagenesis (recrystallization and chertification) of the chalks with which it is conformably interbedded. The volcanigenic sediments are in thin to thick beds and are fine to coarse grained lithic-feldspar-quartz sandstone. The matrix is light colored and commonly calcareous. Pebbles are marl and (or) mudclasts. Volcanigenic beds are tabular and either massive or have distribution grading and predominant Bouma Tb and Tc structures.

We have dated nannofossils and foraminifers in eight marl specimens of the Belvedere Formation near Bogles village from five sites (Figs. 10; Tables 2, 3). Oldest zonal ages in the Bogles Village domain are NP15 to 16, middle Eocene, and the youngest are NP19 to 20, late Eocene (Table 3). The sites of such dates indicate tectonic repetition of Belvedere beds in this domain (Fig. 10). The zonal age of the Bogles Limestone gives a lower stratigraphic bound to the Belvedere of lower NP16 (Fig. 9).

Bedding near Bogles village is partially girdled about a horizontal north-south–trending zonal axis (Fig. 11b). The average dip, about vertical, represents the sheet dip of the western limb of the macroscopic anticline. The variability of dip is due to small folds of possibly two phases: (1) trains of open folds with subhorizontal axial planes of about 4-m wavelength, observed in outcrop; and (2) tight or close steeply dipping minor folds that may be harmonics of the major fold. Although hinges of steeply dipping minor folds have not been seen, their existence is conceived to explain the existence of both east- and west-facing bedding in this domain.

Belvedere southwest. This area contains Belvedere Formation that is on the eastern flank of the dominant anticline where the fold's axial trace trends NNE-SSW (Figs. 10, 11). Southwest of Belvedere, in this area, the beds probably occupy a southeast-dipping upright homocline that is the eastern limb of the dominant anticline. At Belvedere, the structure is more complex, owing to superposed folding and reverse faulting, both of post-Belmont age.

The homocline southwest of Belvedere may be 150 to 200 m thick between the Cherry Hill Basalt and Belmont Formation. The lower and top thirds of this inferred succession are marl-rich whereas the central third contains abundant clastic sediment. The marl-rich zones include variants from nearly pure chalk to mudstone and nests or interbeds of fine- to coarse-grained feldspar-lithic wacke with marly matrix. The marls and clastic rocks both are extensively bioturbated. Some rounded nests of sandstone in marl are probably ball-and-pillow structures.

The clastic sediments in the central third are both volcanigenic and carbonate types. The volcanigenic beds are mainly feldspar-lithic sandstone, which occurs in graded sets with marly tops and in ungraded plane-laminated arenitic layers. Pebbly sandstone occurs abundantly and conglomerate, more rarely. The coarse fragments are pumice and various pelagic intraclasts: chalk, marl, and chert. The matrix of graded sandstones and of the coarser beds includes abundant fine carbonate and clay whose origin we interpret to have been disintegrated marly intraclasts. The pebbly sandstones and conglomerate are notably unsorted and unstratified. The limited exposure prevents assessment whether these beds are channelized or not.

About 200 m south of Belvedere, a marlrich zone of Oligocene age (Fig. 10, site 18) contains widespread intrusive fragmental basalt that is highly vesicular and contains glassy margins. Such basalt is probably correlative with that in Oligocene Belvedere beds in the Windward area where basaltic sediments associated with intrusions indicate they were probably spatter cone feeders in Oligocene time.

We dated marl beds at seven sites and two marl intraclasts from the homocline of the Belvedere southwest domain (Table 3, Fig. 10). In the lower marl, two sites (15, 16, Fig. 10), each probably ≥10 m above the formation base, have zonal ages of NP17 (late middle Eocene). In the intermediate clastic zone, three sites (10, 12, 33) have permissible age ranges of NP17 to 20, NP17 to 20, and NP16 to 20. Two marlclasts (sites 11, 12) in the intermediate zone have a permissible range of NP12 to 20 and a zonal age of NP16? A site in the upper marl (17) gave a zonal range of NP18 to 20. The homoclinal beds, therefore, are entirely Eocene, probably late middle and late Eocene, but conceivably all middle Eocene. We presume that the unexposed basal beds of the homocline have ages in the NP16 zone, as in Bogles village domain to the west.

Also within the Belvedere southwest domain, we sampled two sites in the area of structural complexity near the community of Belvedere (19, 18, Fig. 10). These gave, respectively, zonal ages of NP19 to 20 (late late Eocene) and NP24 to 25 (late Oligocene).

The transition from Eocene to upper Oligocene beds at Belvedere is seemingly rapid, and we postulate this is due to structural juxtaposition rather than a condensed section. The interpretation is based on (1) the probably moderately thick section of lower Oligocene beds in the Belvedere Formation of the Dover area, and (2) the reverse faulting of Belvedere Formation 200 m south of Belvedere (Fig. 10).

Attitudes of locally homoclinal bedding in the Belvedere southwest domain are shown in Figure 10c. Those between the Coulisse and Cherry Hill (solid dots, Fig. 10) dip 30° to 60°, east to southeast, and reflect the upright eastern limb of the macroscopic anticline where its axial trace is NNE-SSW. Near Belvedere, locally homoclinal bedding (open circles, Fig. 10) dips shallowly mainly to the north and south. We interpret the attitudes near Belvedere to reflect the bend in axial trace of the macroscopic anticline from approximately NNE-SSW to EW trend (Fig. 10).

Mesoscopic folds of beds of 5 to 20 m width occur in the Belvedere southwest domain, and a reverse fault that brings Belvedere above Belmont Formation exists 200 m south of Belvedere (Fig. 10). The mesoscopic folds are open to close and upright. Their axes plunge shallowly between south and southeast. They are considered to be harmonics of the major north-trending second phase anticline whose axial trace passes just west of Belvedere because they parallel a second phase minor fold of beds in the Miocene Belmont and Kendeace Formations 400 m south of Belvedere (Fig. 10).

The reverse fault was studied in an excavation. It strikes N47°W, dips 67° northeast, and cuts bedding in the footwall at a high angle. There are no indicators of slip orientation. The movement occurred by fault plane sliding under brittle conditions.

Dover. The Dover area (Fig. 11) extends from just south of Windward village west to near Belvedere. The axial trace of the macroscopic anticline is approximately east-northeast in this area (Fig. 10).

The rocks are generally well bedded successions of marl and sandy turbidites. They are probably higher stratigraphically than most of the Belvedere Formation in the Bogles and Belvedere southwest areas and differ from rocks in those areas by their much smaller content of volcanigenic gravel and greater content of larger benthic forams.

The turbidites are tabular and thin to thick bedded (up to 1.5 m). Their volcanigenic particles are plagioclase, feldspar porphyry, devitrified glass, chloritic aphanite, and clinopyroxene, and they also include carbonate skeletal grains and intraclasts of mudstone and marl. The marly tops of the turbidites and marl interbeds contain nonmixed-layer smectite (Fig. 12), together with biogenic carbonate. Coarse skeletal particles are whole larger forams of the same species that occur in the Belvedere Formation in the Windward area (Fig. 10; Robinson and Jung, 1972; Westercamp and others, 1985). The forams, however, are here mixed with volcanigenic grains in sandstones whereas they are mainly in homogeneous foram microconglomerates at Windward. Bouma zonations in the turbidites are Tae, Tabe, and Tbe.

In the southern Dover area at Meldrum, the Belvedere Formation includes volcanigenic beds of unusual character that occupy a restricted stratigraphic interval. They are laminated and thin-bedded, graded and ungraded basaltic sandstone and mudstone sets. They are noncalcareous and markedly more mature diagenetically, apparently by devitrification of glass, than other Belvedere rocks near Dover. We suspect such beds are a succession of ashfalls partly reworked by bottom currents because of the abundant graded laminae and the absence of skeletal particles.

The Belvedere Formation in the Dover area has been dated at eight sites (Tables 2, 3): five from our work, two from Robinson and Jung (1972; their 10816 and 10815, Fig. 10), and one from Westercamp and others (1985, their Cari 16 site near Dover). The oldest date is at site 21, in the eroded core of the macroscopic anticline, where the permissible age is NP17 to 19 (late middle–late Eocene), assuming *D. binodosus* is resedimented. Other sites that are in presumably higher beds in the

TABLE 3. NANNOFLORA IDENTIFIED IN OLDER ROCKS OF CARRIACOU*

Unit Site	Domain	Sample	Abundance	Preservation	Sphenolithus distentus	Zygrhablithus bijugatus	Sphenolithus furcatolithoides	S. tribulosus	Cyclicargolithus abisectus	Chiasmolithus solitus	Sphenolithus dissimilis	S. ciperoensis	Helicosphaera salebrosa	Discoaster binodosus	Sphenolithus spiniger	Prinsiaceae (small)
Cherry Hill Basalt																
14		90-46a	f	vP	-	-	f	-	-	+	-	-	-	-	+	-
14		90-46b	R	vP	-	-	f	-	-	-	-	-	-	-	-	-
38		10-23-5m2	vR	vP	-	-	-	-	-	-	-	-	-	-	-	-
Bogles Limestone																
13		90-9	R	vP	-	-	-	-	-	-	-	-	-	-	-	-
34		10-11-1m3	f	P	-	-	f	-	-	-	-	-	-	-	-	-
Belvedere Formation																
26	3	Ca5W7	+	vP	-	-	-	-	-	-	-	-	-	-	-	+
23	3	CarW9	+	P	-	-	-	-	-	-	-	-	-	-	-	+
22	3	CarW12	+	P	-	-	-	-	-	-	-	-	-	-	-	+
32	3	87-15	Cf	Pm	-	-	-	-	-	-	-	-	+	+	-	-
18	2	87-24	R	P	-	-	-	-	-	-	R	R	-	-	-	-
15	2	90-35	A	Pm	-	-	-	-	-	-	-	-	+	-	-	-
16	2	90-36c	A	Pm	-	-	-	-	-	-	-	-	+	-	-	-
8	1	90-37b	R	P	-	-	R	-	-	R	-	-	-	-	-	-
9	1	90-43	f	P	-	-	f	-	-	-	-	-	-	-	-	-
5	5	90-45	Cf	Pv	f	-	-	f	f	-	-	-	-	-	-	-
20	3	90-47	Cf	P	f	-	-	f	f	-	-	-	f	-	-	-
19	2	90-48	f	Pm	-	-	-	-	-	-	-	-	-	-	-	-
17	2	9-16-4pl	A	Pm	-	-	-	-	-	-	-	-	-	-	-	C
22	3	90-50	Ca	M	-	-	-	-	-	-	-	-	-	-	-	-
7	1	9-6-13p1	vR	P	-	-	-	-	-	-	-	-	-	-	-	R
7	1	9-6-13p7	R	P	-	?	-	-	-	-	-	-	-	-	-	R
12	2	9-4-4i*	A	Pm	-	-	-	-	-	-	-	-	-	-	+	-
10	2	9-6-10p1	Cf	Pm	-	-	-	-	-	-	-	-	-	-	?	-
11	2	9-4-21si*	vR	P	-	-	-	-	-	-	-	-	-	-	-	R
36	1	9-6-13p2	-	-	-	-	2	-	-	-	-	-	-	-	-	-
36	1	9-6-13p3	-	-	-	-	-	-	-	-	-	-	-	-	-	-
36	1	9-6-13p4	-	-	-	-	-	-	-	-	-	-	-	-	-	-
36	1	9-6-13p5	-	-	-	-	+	-	-	-	-	-	-	-	-	-
35	1	9-7-3p1	-	-	-	-	+	-	-	+	-	-	-	-	-	-
33	2	10-18-3m2	-	-	-	-	-	-	-	-	-	-	-	-	-	-
37	5	10-17-3ls1	A	-	-	-	-	-	-	-	-	-	+	-	-	-
12	2	9-4-4g	Ca	M	-	-	-	-	-	-	-	-	-	-	-	-
25	3	90-55	C	Pm	-	-	-	-	-	-	-	-	-	-	-	-
16	2	10-9-7m2	C	Pm	-	-	-	-	-	-	-	-	-	-	-	-

anticlinal flanks give NP16 to 19 (late Eocene), NP20 to 21, NP23 to 24 (late early–early late Oligocene). NP23 to 25 (late early–late Oligocene), and two dates of NP24 (early late Oligocene; Fig. 10, Table 3). The exposure is so meager and the structure complex enough that we are uncertain of the relative stratigraphic positions of any of these sites.

Structures of the Belvedere Formation in the Dover and Windward domains are alike and shown together in Figure 11d. They comprise two sets of folds. The first set includes macroscopic open or close broad (>100 m) folds with east-northeast-trending axial trace and subhorizontal axes. These folds are indicated by the east-northeast-trending zonal girdle of locally homoclinal beds (Fig. 11d, excepting three attitudes in the southwest quadrant) and by several reversals in dip of local homoclines (Fig. 10). Such folds are harmonics to the macroscopic anticline of the Belvedere Formation where its axial trace trends ENE-WSW. The second set is narrow (5 to 20 m width) mesoscopic open and close upright folds with axial traces and axes trending generally northwest to southeast (Fig. 11d). The second set is interpreted to be the younger and to be superposed on the macroscopic anticline because of the similarity of orientation to folds of the Miocene Belmont Formation near Belvedere.

Windward. In the Windward area, the Belvedere Formation consists of carbonate-rich clastic rocks and marl in a sequence thicker than 100 m. Calcarenites include thick and thin, massive or plane-laminated beds; some are clearly graded throughout the bed whereas others are ungraded or have only abrupt grading to wacke in the top few centimeters. Calcarenite beds with coarse-

TABLE 3. NANNOFLORA IDENTIFIED IN OLDER ROCKS OF CARRIACOU* (continued)

Unit Site	Domain	Sample	Abundance	Preservation	Sphenolithus distentus	Zygrhablithus bijugantus	Sphenolithus furcatolithoides	S. tribulosus	Cyclicargolithus abisectus	Chiasmolithus solitus	Sphenolithus dissimilis	S. ciperoensis	Helicosphaera salebrosa	Discoaster binodosus	Sphenolithus spiniger	Prinsiaceae (small)
Anse La Roche Formation																
4		90-28	Cf	P	-	-	-	-	-	-	-	-	-	-	-	-
1		Ca1	Rf	Pm	-	-	-	-	-	-	-	-	-	-	-	R
1		Ca3	+	vP	-	-	-	-	-	-	-	-	-	-	-	+
2		Car5a	Cf	Mp	-	-	-	-	-	-	-	-	-	-	-	f
3		Car6f1	Ca	Pm	-	-	-	-	-	-	-	-	-	-	-	Ca
3		Car6f2	Cf	Pm	-	-	-	-	-	-	-	-	-	-	-	Cf
3		Car6f3	+	P	-	-	-	-	-	-	-	-	-	-	-	+
3		Car6f4	+	Pm	-	-	-	-	-	-	-	-	-	-	-	+
3		Car6f10	+	P	-	-	-	-	-	-	-	-	-	-	-	+
1		87-26a*	vR	P	-	-	-	-	-	-	-	-	-	-	-	-
1		87-26b*	vR	vP	-	-	-	-	-	-	-	-	-	-	-	-
1		87-26c*	vR	P	-	-	-	-	-	-	-	-	-	-	-	R
6		9-3-2	vR	P	-	-	-	-	-	-	-	-	-	-	-	R
6		9-2-8p*	vR	P	-	-	-	-	-	-	-	-	-	-	-	R
Belmont Formation B																
Fig. 15		87-2	-	-	-	-	-	-	-	-	-	-	-	-	-	-
Kendeace Formation																
Fig. 15		CarK2	+	P	-	-	-	-	-	-	-	-	-	-	-	-
Carriacou Formation																
Fig. 15		CarK1	Rf	P	-	-	-	-	-	-	-	-	-	-	-	-
Fig. 15		CarK3	+	P	-	-	-	-	-	-	-	-	-	-	-	-
Fig. 15		CarK5	Cf	Pm	-	-	-	-	-	-	-	-	-	-	-	-
Fig. 15		CarK8	Cf	P	-	-	-	-	-	-	-	-	-	-	-	-
Fig. 7		Carmp1	Cf	Pm	-	-	-	-	-	-	-	-	-	-	-	-
Fig. 7		Carmp2	+	P	-	-	-	-	-	-	-	-	-	-	-	-

Abundance Code
A = abundant
Ca = common to abundant
C = common
Cf = common to few
f = few
RF = rare to few
R = rare
vR = very rare
# = very very rare
+ = very very very rare

1, 2, ... = number of individuals

Preservation Code
M = moderate
Pm = poor to moderate
P = poor
vP = very poor
? = identification uncertain

*All identifications by K. v. S. Perch-Nielsen. Sample sites on Figure 10 except where noted; domains of Belvedere Formation shown on Figure 11. Asterisked samples are rock clasts.

grained basal zones have scoured bottoms. The conglomerate beds have touching, well-oriented platy neritic benthic large foraminifers and occasional outsize micrite clasts. Such conglomerates are much like those of the Anse La Roche Formation except that they are constituted by entirely younger larger benthic foram species (Robinson and Jung, 1972). This series of skeletal-clastic limestones is interpreted to have arisen by sediment-gravity flows from a shallow-marine source region to a basinal environment. The evidence is the existence of planktic foram-bearing micritic clasts in the coarse beds and the lack of abrasion of most larger benthic forams.

Within the beds at Windward are sporadic small dikes and plugs of aphyric or olivine basalt. These grade out to envelopes of breccia composed of mixed fragments of basalt, skeletal carbonate, and micrite; the breccia is cemented by carbonate. On the fringes of some breccia bodies and away from exposed basalt, stratified microbreccia or sandstone of mixed skeletal and basaltic particles exist. The stratification is due to varying ratios of the two particle types and grain size changes. From such features, extrusion of basalt probably occurred in the Belvedere Formation basin syndepositionally in late Oligocene time.

We obtained no dates from the Windward area. Wester-

TABLE 3. NANNOFLORA IDENTIFIED IN OLDER ROCKS OF CARRIACOU* (continued)

Unit Site	Domain	Sample	Coccolithus eopelagius	Cyclicargolithus floridanus	Coccolithus pelagicus	Dictyococcites cf. D. antarcticus	D. bisecta	Ericsonia formosa	Reticulofenestra umbilicus	Chiasmolithus (small)	Pontosphaera multipora	Calcidiscus leptoporus	Braarudosphaera bigelowii	Thoracosphaera sp.	Ericsonia obruta
Cherry Hill Basalt															
14		90-46a	-	-	-	-	-	-	R	-	-	-	-	-	-
14		90-46b	-	-	-	-	-	-	+	-	-	-	-	-	-
38		10-23-5m1	-	R	-	-	-	-	-	-	-	-	-	-	-
Bogles Limestone															
13		90-9	-	-	-	-	-	-	-	-	-	-	-	-	-
34		10-111m3	-	-	-	-	-	-	-	-	-	-	-	-	-
Belvedere Formation															
26	3	Ca5W7	-	-	-	-	-	-	-	-	-	-	-	-	-
23	3	CarW9	-	-	-	-	-	-	-	-	-	-	-	-	-
22	3	CarW12	+	+	+	-	+	-	-	+	-	-	-	-	1
32	3	87-15	-	-	-	-	+	+	+	+	-	-	-	-	-
18	2	87-24	-	-	-	-	R	-	-	-	-	-	-	-	-
15	2	90-35	-	-	-	-	+	-	+	-	-	-	-	-	-
16	2	9036c	-	-	-	-	+	-	+	-	-	-	-	-	-
8	1	90-37b	-	-	-	-	-	-	+	-	-	-	-	-	-
9	1	90-43	-	-	-	-	-	-	-	-	-	-	-	-	-
5	5	90-45	-	-	-	-	-	-	-	-	-	-	-	-	-
20	3	90-47	-	-	-	-	-	-	-	-	-	-	-	-	-
19	2	90-48	-	-	-	-	-	-	-	-	-	-	-	-	-
17	2	9-16-4l	R	Cf	R	-	R	3	3	1	-	-	-	R	-
22	3	90-50	-	-	-	-	-	-	-	-	-	-	-	-	-
7	1	9-6-13p1	-	R	R	-	-	-	-	-	-	-	-	-	-
7	1	9-6-13p7	-	R	R	-	-	-	-	-	-	-	-	-	-
12	2	9-4-4i*	-	-	-	-	+	+	+	+	-	-	-	-	-
10	2	9-6-10p1	-	-	-	-	?	?	?	-	-	-	-	-	-
11	2	9-4-21si*	R	-	R	-	-	R	-	-	-	-	-	-	-
36	1	9-6-13p2	-	-	-	-	-	-	-	-	-	-	-	-	-
36	1	9-6-13p3	-	-	-	-	-	-	-	-	-	-	-	-	-
36	1	9-6-13p4	-	-	-	-	-	-	-	-	-	-	-	-	-
36	1	9-6-13p5	-	-	-	-	-	-	-	-	-	-	-	-	-
35	1	9-7-3p1	-	-	-	-	-	-	-	-	-	-	-	-	-
33	2	10-18-3m2	-	-	-	-	-	-	+	-	-	-	-	-	-
37	5	10-17-3ls1	-	-	-	-	-	-	-	-	-	-	-	-	-
12	2	9-4-4g	-	-	-	-	-	-	-	-	-	-	-	-	-
25	3	90-55	-	-	-	-	-	-	-	-	-	-	-	-	-
16	2	10-9-7m2	-	-	-	-	-	-	-	-	-	-	-	-	-

camp and others (1985), however, dated site Cari 15 (Fig. 10) with 11 planktic species whose zonal age is NP24, early late Oligocene. The only discordant species in Cari 15 is *Globorotalia obesa* whose first appearance is early Miocene (*C. dissimilis* zone) according to Bolli and others (1985). Owing to the overlap of so many species in the late Oligocene, however, we tentatively consider the maximum age of *G. obesa* poorly established rather than the fauna having been resedimented in the Miocene.

Larger foraminifer species in the Windward area reported by Robinson and Jung (1972) and Westercamp and others (1985) with one exception have first appearance in the Oligocene, as young as NP24, according to species ranges from Butterlin (1981), Andreieff (1985), and J. P. Beckmann (written communication, 1986). There is one exception, *Miogypsina gunteri*, which began in early Miocene time according to Butterlin (1981). Assuming the age ranges of the *G. obesa* and *M. gunteri* are poorly known, the strata of the Windward area include lower upper Oligocene beds (NP24).

Meager exposure prevents direct assessment of stratigraphic relationships between Belvedere beds in the Dover and Windward areas. The two areas both include lower upper Oligocene beds. The lithology of the two areas differs substantially in the high ratio of volcanigenic to skeletal grains in clastic rocks of the Dover area and vice versa at Windward. The few bedding attitudes in the transition area imply the beds at Windward are the lower (Fig. 10). If so, the carbonate clastic section at Windward

TABLE 3. NANNOFLORA IDENTIFIED IN OLDER ROCKS OF CARRIACOU* (continued)

Unit Site	Domain Sample	Coccolithus eopelagius	Cyclicargolithus floridanus	Coccolithus pelagicus	Dictyococcites cf. D. antarcticus	D. bisecta	Ericsonia formosa	Reticulofenestra umbilicus	Chiasmolithus (small)	Pontosphaera multipora	Calcidiscus leptoporus	Braarudosphaera bigelowii	Thoracosphaera sp.	Ericsonia obruta
Anse La Roche Formation														
4	90-28	-	-	-	-	+	-	+	-	-	-	-	-	-
1	Ca1	-	+	R	+	+	#	+	1	-	-	-	-	-
1	Ca3	-	-	-	1	+	-	-	-	-	-	-	-	-
2	Car5a	+	-	-	-	+	#	R	-	-	-	-	-	+
3	Car6f1	+	R	-	R	-	-	-	-	-	-	-	-	+
3	Car6f2	+	R	-	-	-	-	-	-	-	-	-	-	1
3	Car6f3	-	+	-	-	-	-	-	-	-	-	-	-	-
3	Car6f4	-	-	-	-	-	-	-	-	-	-	-	-	-
3	Car6f10	-	-	-	-	-	-	-	-	-	-	-	-	-
1	87-26a*	-	R	-	-	R	-	-	-	-	-	-	-	-
1	87-26b*	-	R	-	-	R	-	-	-	-	-	-	-	-
1	87-26c*	-	-	-	-	R	-	-	-	-	-	-	-	-
6	9-3-2	-	-	R	-	R	-	-	-	-	-	-	-	-
6	9-2-8p*	-	-	R	-	R	-	-	-	-	-	-	-	-
Belmont Formation B														
Fig. 15	87-2	-	R	-	R	-	-	-	-	-	-	-	-	-
Kendeace Formation														
Fig. 15	CarK2	-	+	-	-	1	1	-	-	-	-	-	-	-
Carriacou Formation														
Fig. 15	CarK1	-	Rf	-	-	+	+	-	-	-	-	1	1	-
Fig. 15	CarK3	-	+	-	-	1	-	-	-	-	-	-	-	-
Fig. 15	CarK5	-	fC	-	1	+	+	-	-	-	-	-	-	1
Fig. 15	CarK8	-	fC	-	-	-	+	-	-	-	-	-	-	-
Fig. 7	Carmp1	-	f	-	+	+	+	-	-	-	-	-	-	-
Fig. 7	Carmp2	-	+	-	-	-	-	-	-	-	-	-	-	-

should exist, but is unrecognized, in the Dover area which includes Eocene as well as Oligocene beds. Conceivable relations are as follows: (1) the beds at Windward are younger than those at Dover and beds of the two areas are in contact at an unseen fault; (2) beds at Windward occur unexposed in the Dover area; and (3) beds of the two areas are facies. The first and third ideas are equally probable. The second is doubtful because the section at Windward is thick and the beds are resistant, and their existence should be evident in slope debris. If the beds of the two areas are facies, those at Windward are probably the fill of a major channel that cut through the turbidite fan represented by the beds of the Dover area.

The structure of the Belvedere Formation at Windward is an east-northeast–trending upright anticline with subhorizontal axis (Figs. 10, 11d). Limb dips are as great as 75°. The anticline is colinear with east-northeast–trending folds in the Dover area (Fig. 11d) and is a harmonic of the macroscopic anticline. Thrust faults of small displacement (<2 m) are recognized in good outcrops (e.g., Moule Point, Fig. 10). These have varied orientations, and their relation to folding is uncertain.

High North, west flank. Belvedere Formation is exposed in several gullies on the western flank of High North (Fig. 10) where the formation is unconformable below Belmont Formation and has an inferred thrust contact above the Anse La Roche Formation. The region between these gullies and Bogles and Belvedere is virtually unexposed, and our interpolation therein is highly approximate.

The Belvedere Formation in the gullies consists of (1) marl and mudstone with interbeds of thin-bedded, fine-grained sandstone with mixed volcanigenic and skeletal grains; and (2) foram microconglomerates that are strongly cemented in medium and thick beds, like those of the Windward area.

We dated marl at two sites (Fig. 10; Tables 2, 3): in NP23 to 24 (site 5), late early or early late Oligocene; and NP 25 (site 37), end Oligocene.

The principal structural interpretation from the Belvedere on the west flank of High North is that the thrust which cuts the overturned macroscopic anticline at Bogles village (Fig. 10) probably extends north, unexposed between Belvedere beds of the Bogles area and those west of High North. The thrust may cut progressively higher beds in the hangingwall toward the north as shown in Figure 10. Alternatively, the thrust could be northeast of the Belvedere beds west of High North, and these beds could be in section with those of the Bogles area. A corollary of the alternative scheme is that the Bogles thrust cuts upsection in the hangingwall toward the north.

TABLE 3. NANNOFLORA IDENTIFIED IN OLDER ROCKS OF CARRIACOU* (continued)

Unit Site	Domain	Sample	Discoaster sp (7 arms)	Discoaster sp (6 arms)	Discoaster sp (5 arms)	Discoaster delflandrei	D. damanteus	D. saipanensis	D. barbadensis	Sphenolithus radians	S. predistentus	S. heteromorphus	S. moriformis	Helicosphaera euphratis	H. compacta	H. obliqua/ gertae	H. kamptneri	Helicosphaera sp.
Cherry Hill Basalt																		
14		90-46a	-	-	-	-	-	-	-	-	-	-	-	-	-	-	-	-
14		90-46b	-	-	-	-	-	-	-	-	-	-	-	-	-	-	-	-
38		10-23-5m2	-	-	-	-	-	R	R	R	-	-	R	-	-	-	-	-
Bogles Limestone																		
13		90-9	-	-	-	-	-	+	+	-	-	-	-	-	-	-	-	-
34		10-111m3	-	-	-	-	-	-	-	-	-	-	-	-	-	-	-	-
Belvedere Formation																		
26	3	Ca5W7	-	-	-	-	-	-	-	-	-	-	-	-	-	-	-	-
23	3	CarW9	-	-	-	-	-	-	-	-	-	-	1	-	-	-	-	-
22	3	CarW12	-	-	-	-	-	-	-	-	-	-	+	+	-	-	-	-
32	3	87-15	-	-	-	-	-	+	+	+	-	-	-	-	+	-	-	-
18	2	87-24	-	-	-	-	-	-	-	-	-	-	-	-	-	-	-	-
15	2	90-35	-	-	-	-	-	-	-	+	-	-	-	-	+	-	-	-
16	2	9036c	-	-	-	-	-	-	-	+	-	-	-	-	-	-	-	-
8	1	90-37b	-	-	-	-	-	-	-	-	-	-	-	-	-	-	-	-
9	1	90-43	-	-	-	-	-	-	-	-	-	-	-	-	-	-	-	-
5	5	90-45	-	-	-	-	-	-	-	-	-	-	-	-	-	-	-	-
20	3	90-47	-	-	-	-	-	-	-	-	-	-	-	-	-	-	-	-
19	2	90-48	-	-	-	-	-	-	-	-	-	-	-	-	-	-	-	-
17	2	9-16-4pl	2	3	2	6	-	3	f	1	-	-	Rf	1	2	-	-	1
22	3	90-50	-	-	-	-	-	+	+	-	-	-	-	-	-	-	-	-
7	1	9-6-13p1	-	-	-	-	-	-	?	-	-	-	-	-	-	-	-	-
7	1	9-6-13p7	-	-	-	-	-	-	-	-	-	-	-	-	-	-	-	-
12	2	9-4-4i*	-	-	-	-	-	+	+	-	-	-	+	-	+	-	-	-
10	2	9-6-10p1	-	-	-	-	-	?	?	-	-	-	?	-	?	-	-	-
11	2	9-4-21si*	-	-	-	-	-	-	R	-	-	-	-	-	-	-	-	-
36	1	9-6-13p2	-	-	-	-	-	-	-	-	-	-	-	-	-	-	-	-
36	1	9-6-13p3	-	-	-	-	-	?	?	-	-	-	-	-	-	-	-	-
36	1	9-6-13p4	-	-	-	-	-	?	?	-	-	-	-	-	-	-	-	-
36	1	9-6-13p5	-	-	-	-	-	-	-	-	-	-	-	-	-	-	-	-
35	1	9-7-3p1	-	-	-	-	-	-	-	-	-	-	-	-	-	-	-	-
33	2	10-18-3m2	-	-	-	-	-	+	+	-	-	-	-	-	-	-	-	-
37	5	10-17-3ls1	-	-	-	-	-	-	-	-	-	-	-	-	-	-	-	-
12	2	9-4-4g	-	-	-	-	-	+	+	-	-	-	-	-	-	-	-	-
25	3	90-55	-	-	-	-	-	+	+	-	-	-	-	-	-	-	-	-
16	2	10-9-7m2	-	-	-	-	-	-	-	-	-	-	-	-	-	-	-	-

Discussion. The Belvedere Formation contains a restricted range of lithotypes, carbonate-rich pelagic sediment, and sandy volcanigenic skeletal turbidite that we believe unifies the formation as a stratigraphic succession. Deposition occurred in a relatively uniform environment over a substantial duration of 15 to 20 m.y.—late middle Eocene (NP16) to at least late late Oligocene (NP25).

Zonal ages of the Belvedere Formation range between NP16 to 20 and NP23 to 25 (late middle and late Eocene and late early and late Oligocene; Fig. 9). The permissible age range is NP15 to 25, at least 25 m.y. according to Berggren and others (1985). The conceivable stratigraphic range is NP16 to NN4 (early Miocene), the older bound from the Bogles Limestone and the younger from the Kendeace Formation. The Kendeace is the lowest dated unit that is demonstrably depositional on the Belvedere (Fig. 9).

The onset of deposition of the Belvedere in NP16 time upon the Cherry Hill Basalt and, perhaps, the Bogles Limestone is fairly confidently established. It is less certain, however, whether the Belvedere contains a lower Oligocene hiatus or not and whether it contains or did contain beds younger than NP25. Regarding the first question, our dates are permissive of early Oligocene ages at only two sites, and at both, late Oligocene is equally likely. Therefore it is possible that beds of early Oligocene age may be absent or in an unsampled condensed section, or that lower Oligocene beds may be omitted or buried by unrecognized intra-Belvedere structures.

TABLE 3. NANNOFLORA IDENTIFIED IN OLDER ROCKS OF CARRIACOU* (continued)

Unit Site	Domain	Sample	Discoaster sp (7 arms)	Discoaster sp (6 arms)	Discoaster sp (5 arms)	Discoaster delflandrei	D. Adamanteus	D. saipanensis	D. barbadensis	Sphenolithus radians	S. predistentus	S. heteromorphus	S. moriformis	Helicosphaera euphratis	H. compacta	H. obliqua/gertae	H. kamptneri	Helicosphaera sp.
Anse La Roche Formation																		
4		90-28	-	-	-	-	-	-	-	-	-	-	-	-	-	-	-	-
1		Ca1	-	1	#	R	-	-	-	-	-	-	#	-	-	-	-	-
1		Ca3	-	-	-	-	-	-	-	-	-	-	1	-	-	-	-	-
2		Car5a	1	+	-	+	-	#	+	1	-	-	R	-	-	-	-	-
3		Car6f1	-	+	-	-	1	-	-	-	#	-	f	-	#	-	-	-
3		Car6f2	-	+	-	-	-	-	-	-	1	-	f	-	+	-	-	-
3		Car6f3	-	-	-	-	-	-	-	-	-	-	+	-	-	-	-	-
3		Car6f4	-	-	-	-	-	-	-	-	-	-	-	-	-	-	-	-
3		Car6f10	-	-	-	-	-	-	-	-	-	-	1	-	-	-	-	-
1		87-26a*	R	R	R	-	-	-	-	-	-	-	R	-	-	-	-	-
1		87-26b*	R	R	R	-	-	-	-	-	-	-	R	-	-	-	-	-
1		87-26c*	-	-	-	-	-	-	-	-	-	-	-	-	-	-	-	-
6		9-3-2	-	-	-	-	-	-	R	-	-	-	-	-	-	-	-	-
6		9-2-8p*	-	-	-	-	-	-	-	-	-	-	-	-	-	-	-	-
Belmont Formation B																		
Fig. 15		87-2	-	-	-	-	-	-	-	-	-	-	-	R	-	-	-	-
Kendeace Formation																		
Fig. 15		CarK2	-	-	-	-	-	-	-	-	-	-	1	-	-	-	-	1
Carriacou Formation																		
Fig. 15		CarK1	-	-	-	+	-	-	-	-	-	-	1	-	-	-	#	#
Fig. 15		CarK3	-	-	-	-	-	-	-	-	-	-	-	-	-	-	-	-
Fig. 15		CarK5	-	-	-	f	-	f	f	-	-	#	+	-	-	1	+	+
Fig. 15		CarK8	-	-	-	-	-	-	-	-	-	-	-	-	-	-	-	-
Fig. 7		Camp1	-	-	-	-	-	-	-	-	-	+	+	-	-	-	1	1
Fig. 7		Camp2	-	-	-	-	-	-	-	-	-	-	-	-	-	-	1	-

The existence of beds younger than NP25 in the Belvedere Formation can be hinted at from the disparity in ages of dated Belvedere sites nearest the unconformable base of the Belmont Formation (Fig. 10). The late Eocene age near the upper contact of the Belvedere near the Coulisse (Fig. 10) implies that the Belvedere was variably eroded prior to Belmont deposition and that beds younger than NP25 may have originally capped the Belvedere.

The Belvedere Formation accumulated in a basin below wave base for its entire duration. A carbonate-bearing pelagic downfall occurred almost steadily together with episodic incursions of turbidites of volcanigenic and (or) carbonate skeletal grains. Ash and pumice falls may have contributed sediment. The pelagic background indicates depositional depth above the carbonate compensation depth (CCD). The mudstones of the Belvedere probably record high influx of volcanigenic mud rather than sub-CCD conditions because of the absence of radiolarians in such deposits. The late Eocene CCD was shallow, about 3.5 km, in tropical zones but markedly deeper (6 km) in the Oligocene (Kennett, 1978); we do not recognize compositional change in Belvedere rocks related to the Eocene-Oligocene transition and therefore presume the depth was shallower than 3.5 km throughout its duration.

The volcanigenic sediment is mainly or wholly arc derived as indicated by the petrography of grains: coarsely porphyritic, partly pyroxene and hornblende bearing, partly quartz bearing, and a diversity of lithotypes. The source region was certainly shallow marine and perhaps, partly subaerial, at least in Oligocene time, as indicated by the persistent mixing of neritic and (or) littoral benthic fauna in the turbidites. The turbidites are probably mainly outer fan or fan fringe deposits by virtue of the tabularity, grain size, and Bouma zonations of most of them. The widespread content in thick Belvedere turbidites of marly intraclasts, however, indicates passage of the sediment-gravity flows through channels cutting contemporary pelagic sediments before reaching the site of deposition. There is no evidence for transport directions in Belvedere turbidites.

The upper Oligocene beds of the Belvedere Formation at Windward (Fig. 10) contain a variation on the depositional theme just described wherein the sediment-gravity flows were composed almost exclusively of skeletal grains. As noted, it is unclear whether the beds at Windward are the youngest preserved in the Belvedere or whether they are an upper Oligocene facies, perhaps a large channel fill, of the marl-volcanigenic turbidite succession. If they are youngest, the beds at Windward may record a cessation of arc volcanism and transition to a car-

TABLE 3. NANNOFLORA IDENTIFIED IN OLDER ROCKS OF CARRIACOU* (continued)

Unit Site	Domain	Sample	Eiffellithus turriseiffelii	Micrantholithus sp.	Cribocentrum reticulatum	Campylosphaera oronocycles	Coronocyclus prionion	Pedinocyclus larvalis	Calcidiscus protoannulus	Cruciplacolithus sp.	Pseudotriquetrorhabulus inversus	Sphenolithus intercalaris	Isthmolithus recurvus	Helicosphaera heezenii	Sphenolithus sp	Chiasmolithus consuetus
Cherry Hill Basalt																
14		90-46a	-	-	-	-	-	-	-	-	-	-	-	-	-	-
14		90-46b	-	-	-	-	-	-	-	-	-	-	-	-	-	-
38		10-23-5m2	-	-	-	-	-	-	-	-	-	-	-	-	-	-
Bogles Limestone																
13		90-9	-	-	-	-	-	-	-	-	f	-	-	-	-	-
34		10-111m3	-	-	-	-	-	-	-	-	-	-	-	-	-	-
Belvedere Formation																
26	3	Ca5W7	-	-	-	-	-	-	-	-	-	-	-	-	-	-
23	3	CarW9	-	-	-	-	-	-	-	-	-	-	-	-	-	-
22	3	CarW12	-	-	-	-	-	-	-	-	-	-	-	-	-	-
32	3	87-15	-	-	?	-	-	-	-	-	-	-	-	-	-	-
18	2	87-24	-	-	-	-	-	-	-	-	-	-	-	-	-	-
15	2	90-35	-	-	-	-	-	-	-	-	-	+	+	-	-	-
16	2	9036c	-	-	-	-	-	-	-	-	-	+	+	-	-	-
8	1	90-37b	-	-	-	-	-	-	-	-	-	-	-	-	-	-
9	1	90-43	-	-	-	-	-	-	-	-	-	-	-	-	-	-
5	5	90-45	-	-	-	-	-	-	-	-	-	-	-	-	-	-
20	3	90-47	-	-	-	-	-	-	-	-	-	-	-	-	-	-
19	2	90-48	-	-	-	-	-	-	-	-	-	-	-	R	-	-
17	2	9-16-4pl	-	-	-	-	-	-	R	1	-	-	-	-	-	-
22	3	90-50	-	-	+	-	-	-	-	-	-	-	-	-	-	-
7	1	9-6-13p1	-	-	-	-	-	-	-	-	-	-	-	-	-	-
7	1	9-6-13p7	-	-	-	-	-	-	-	-	-	-	-	-	R	-
12	2	9-4-4i*	-	-	-	+	+	-	-	-	-	-	-	+	+	+
10	2	9-6-10p1	-	-	-	-	-	-	-	-	-	-	-	-	-	-
11	2	9-4-21si*	-	-	-	-	-	-	-	-	-	-	-	-	-	-
36	1	9-6-13p2	-	-	-	-	-	-	-	-	-	-	-	-	-	-
36	1	9-6-13p3	-	-	-	-	-	-	-	-	-	-	-	-	-	-
36	1	9-6-13p4	-	-	-	-	-	-	-	-	-	-	-	-	-	-
36	1	9-6-13p5	-	-	-	-	-	-	-	-	-	-	-	-	-	-
35	1	9-7-3p1	-	-	-	-	-	-	-	-	-	-	-	-	-	-
33	2	10-18-3m2	-	-	?	-	-	-	-	-	-	-	-	-	-	-
37	5	10-17-3ls1	-	-	-	-	-	-	-	-	-	-	-	-	-	-
12	2	9-4-4g	-	-	-	-	-	-	-	-	-	-	-	-	-	-
25	3	90-55	-	-	+	-	-	-	-	-	-	-	-	-	-	-
16	2	10-9-7m2	-	-	+	-	-	-	-	-	+	-	-	-	-	-

bonate platform at the sediment source. If they are facies, they may record the development of a carbonate platform whose sediment had direct access to the late Oligocene Belvedere basin, bypassing conduits of volcanic sediment.

A local basaltic vent system breached Belvedere strata in late Oligocene time and produced minor volcanic sediment, perhaps from spatter cones, but no recognized lava. We do not know whether such basalt is of arc or intraplate spreading origin.

Anse La Roche Formation. Introduction. The Anse La Roche Formation crops out in western northern Carriacou between Hillsborough and Gun Point (Figs. 7, 10) below the Bogles thrust and unconformable cover of Miocene Belmont Formation. Its base is unexposed. It was named by Robinson and Jung (1972) for generally coarse volcanigenic and biogenic clastic sediment of sediment-gravity flow origin and minor intercalated hemipelagic marl and mudstone. Fossil dates at sites indicate the Anse La Roche contains upper Eocene and lower Oligocene beds. Lack of outcrop and structural complexity preclude resolution of a clear stratigraphy and thickness for the formation as a whole. The thickest continuous exposure is about 300 m at Sparrow Bay, which is described separately below (Figs. 10, 13).

The Anse La Roche Formation is the structurally lowest unit on Carriacou (Figs. 8, 10) and forms the autochthon to the Bogles thrust. Its predominant structure is a train of upright close macroscopic folds with east-northeast-trending axial traces (Fig. 10). Near Bogles village (Fig. 10), the Anse La Roche has anom-

TABLE 3. NANNOFLORA IDENTIFIED IN OLDER ROCKS OF CARRIACOU* (continued)

Unit Site	Domain	Sample	Eiffellithus turriseiffelii	Micrantholithus sp.	Cribocentrum reticulatum	Campyloshaera oronocycles	Coronocyclus prionion	Pedinocyclus larvalis	Calcidiscus protoannulus	Cruciplacolithus sp.	Pseudotriquetrorhabulus inversus	Sphenolithus intercalaris	Isthmolithus recurvus	Helicosphaera heezenii	Sphenolithus sp	Chiasmolithus consuetus
Anse La Roche Formation																
4		90-28	-	-	+	-	-	-	-	-	-	-	-	-	-	-
1		Ca1	-	-	-	-	-	-	-	-	-	-	-	-	-	-
1		Ca3	-	-	-	-	-	-	-	-	-	-	-	-	-	-
2		Car5a	-	-	-	-	-	-	-	-	-	-	-	-	-	-
3		Car6f1	-	-	-	-	-	-	-	-	-	-	-	-	-	-
3		Car6f2	-	-	-	-	-	-	-	-	-	-	-	-	-	-
3		Car6f3	-	-	-	-	-	-	-	-	-	-	-	-	-	-
3		Car6f4	-	-	-	-	-	-	-	-	-	-	-	-	-	-
3		Car6f10	-	-	-	-	-	-	-	-	-	-	-	-	-	-
1		87-26a*	-	-	-	-	-	-	-	-	-	-	-	-	-	-
1		87-26b*	-	-	-	-	-	-	-	-	-	-	-	-	-	-
1		87-26c*	-	-	-	-	-	-	-	-	-	-	-	-	-	-
6		9-3-2	-	-	-	-	-	-	-	-	-	-	-	-	-	-
6		9-2-8p*	-	-	-	-	-	-	-	-	-	-	-	-	-	-
Belmont Formation B																
Fig .15		87-2	-	-	-	-	-	-	-	-	-	-	-	-	-	-
Kendeace Formation																
Fig. 15		CarK2	-	-	-	-	-	-	-	-	-	-	-	-	-	-
Carriacou Formation																
Fig. 15		CarK1	1	1	-	-	-	-	-	-	-	-	-	-	-	-
Fig. 15		CarK3	-	-	-	-	-	-	-	-	-	-	-	-	-	-
Fig. 15		CarK5	-	-	-	-	-	-	-	-	-	-	-	-	-	-
Fig. 15		CarK8	-	-	-	-	-	-	-	-	-	-	-	-	-	-
Fig. 7		Carmp1	-	-	-	-	-	-	-	-	-	-	-	-	-	-
Fig. 7		Carmp2	-	-	-	-	-	-	-	-	-	-	-	-	-	-

alous structure: steeply east dipping overturned beds, which we interpret to be a deformational overprint on the east-northeast-trending foldtrain in the footwall ramp of the Bogles thrust. The Anse La Roche beds are moderately faulted and frequently cut by sills and nearly vertical dikes of basalt of known and presumed Neogene age.

Rocks. Exposed strata of the Anse La Roche Formation are chiefly volcanigenic clastics, consisting in diminishing proportion of pebbly sandstone, coarse-, medium-, and fine-grained sandstone, and conglomerate. Beds of mudstone and marly mudstone are minor but pervasive constituents. The clastic beds include minor to locally major quantities of skeletal particles, but the matrix is typically noncalcareous. Other principal lithic characteristics are thick bedding, high proportion of detrital feldspar and light-colored volcanilithics, and green or colorless muddy intraclasts.

The strata are a succession of sediment-gravity flows that are mainly inner- and midfan facies. There is a strong correlation between grain size and thickness of depositional units, each of which is a product of a discrete flow or more rarely, an amalgamated sandstone from a series of scouring flows.

Thick units, 1 to >25 m thick, are either massive or graded. Massive units are mainly pebbly sandstone, but occasionally conglomerate and coarse-grained sandstone, which have no stratification, sorting, or preferred particle orientation. Graded units contain in sequence portions or all of the following: massive pebbly sandstone or conglomerate; pebbly sandstone with upward fining of pebbles, fair to good pebble orientation, and occasional rude plane lamination; coarse- to medium-grained sandstone that is plane and cross laminated and upward fining, and that commonly contains mudclasts; fine-grained sandstone that grades up to a mudstone and (or) marl top. Among graded units, top absent (Tab) or full gradation (Tabce) are more prevalent than base absent (Tbce or Tce). Bases of thick units are both channeled and apparently planar within the limited outcrop area.

Thin clastic units are tabular turbidites 0.1 to 1 m thick in sets ≥10 m thick. These are generally sandrich but top present. Their internal zonations are Tae, Tabe, Tbe, and Tbce.

The hemipelagic layers are commonly thin (<10 cm) discontinuous tops to graded units, but they are occasionally thicker, as much as 0.5 m. They are structureless; comprise mudstone, marl, and calcareous sandy mudstone; and are turquoise or apple green where fresh and soft.

The gravel fraction of the thick units comprises hard igneous and skeletal lithoclasts and soft intraclasts of pelagic and turbiditic mudstone. Igneous lithoclasts of a given bed are as coarse as 1 m

TABLE 3. NANNOFLORA IDENTIFIED IN OLDER ROCKS OF CARRIACOU* (continued)

Unit Site	Domain	Sample	Sphenolithus cf S. obtusus	Discoaster gemmeus	Helicosphaera trumpyii	Chiasmolithus nitidus	Sphenolithus grandis	Campylosphaera dela	Nannoflora Zonal Range
Cherry Hill Basalt									
14		90-46a	-	-	-	+	+	-	Lower NP16
14		90-46b	-	-	-	-	-	-	Lower NP16
38		10-23-5m2	-	-	-	-	-	-	NP11-16
Bogles Limestone									
13		90-9	-	-	-	-	-	-	NP16-17
34		10-111m3	-	-	-	-	-	-	NP15-16
Belevedere Formation									
26	3	Ca5W7	-	-	-	-	-	-	
23	3	CarW9	-	-	-	-	-	-	NP12-NN9
22	3	CarW12	-	-	-	-	-	-	NP18-25
32	3	87-15	-	-	-	-	-	-	NP16-17
18	2	87-24	-	-	-	-	-	-	NP24-25
15	2	90-35	-	-	-	-	-	-	NP17
16	2	9036c	-	-	-	-	-	-	NP17
8	1	90-376	-	-	-	-	-	-	NP16
9	1	90-43	-	-	-	-	-	-	NP15-16
5	5	90-45	-	-	-	-	-	-	NP23-24
20	3	90-47	-	-	-	-	-	-	NP23-24
19	2	90-48	-	-	-	-	-	-	NP19-20
17	2	9-16-4pl	-	-	-	-	-	-	NP18-20
22	3	90-50	-	-	-	-	-	-	NP16-20
7	1	9-6-13p1	-	-	-	-	-	-	NP16-20
7	1	9-6-13p7	R	?	-	-	-	-	NP16-18
12	2	9-4-4i*	-	-	-	-	-	+	NP17-20
10	2	9-6-10p1	-	-	-	-	-	-	NP17-20
11	2	9-4-21si*	-	-	-	-	-	-	NP12-20
36	1	9-6-13p2	-	-	-	-	-	-	NP15-16?
36	1	9-6-13p3	-	?	-	-	-	-	NP16?
36	1	9-6-13p4	-	?	-	-	-	-	NP16?
36	1	9-6-13p5	-	-	-	-	-	-	NP15-16
35	1	9-7-3p1	-	-	-	-	-	-	NP15-16
33	2	10-18-3m2	-	-	-	-	-	-	NP16-20
37	5	10-17-3ls1	-	-	+	-	-	-	NP25
12	2	9-4-4g	-	+	-	-	-	-	NP16?
25	3	90-55	-	-	-	-	-	-	NP17-18
16	2	10-9-7m2	-	-	-	-	-	-	NP17-18

diameter and commonly vary from angular to subrounded. In a few beds, they are entirely angular. Lithoclasts are always polymict and mainly feldspar porphyry, feldspar-pyroxene porphyry, basalt, hornblende porphyry, and diorite, which show no presedimentation alteration or metamorphism. Skeletal lithoclasts are commonly larger benthic foraminifers (Cole, 1958; Robinson and Jung, 1972; Westercamp and others, 1985) that are little abraded or broken and algal fragments. The skeletal granules and fine pebbles are minor to absent in most layers but compose 50 to 100% of the gravel in some thick units. The intraclasts are as coarse 2 m in diameter and are angular and twisted. They occur both in tabular concentrations, typically at or near Ta-Tb transitions in graded units, and dispersed in massive and graded pebbly sandstones. The intraclasts vary from smectitic mudstone to marl to laminated fine sandy mudstone. This lithic range is like that of tops of graded units in the Anse La Roche, but the intraclasts are generally more calcareous. This may indicate either that the intraclasts came from a different environment upcurrent or that the more calcareous sediment preferentially escaped disaggregation in the sediment-gravity flows.

The sand fraction in thick and thin units consists predominantly of plagioclase and feldspathic volcanilithics. Pyroxene, hornblende, and skeletal fragments are pervasive minor constituents. The matrix of finer sandstone and the mudstone and marl tops of graded units consist of smectite with only slight content of illite (Fig. 12).

Section at northern Sparrow Bay. The 300 m thick sequence of strata in northern Sparrow Bay (Fig. 10) provides a good

TABLE 3. NANNOFLORA IDENTIFIED IN OLDER ROCKS OF CARRIACOU* (continued)

Unit Site	Domain	Sample	Sphenolithus cf S. obtusus	Discoaster gemmeus	Helicosphaera trumpyii	Chiasmolithus nitidus	Sphenolithus grandis	Campylosphaera dela	Nannoflora Zonal Range
Anse La Roche Formation									
4		90-28	-	-	-	-	-	-	NP16-19
1		Ca1	-	-	-	-	-	-	NP17-21
1		Ca3	-	-	-	-	-	-	NP16-21
2		Car5a	-	-	-	-	-	-	NP17-20
3		Car6f1	-	-	-	-	-	-	NP17-23
3		Car6f2	-	-	-	-	-	-	NP17-23
3		Car6f3	-	-	-	-	-	-	NP16-NN7
3		Car6f4	-	-	-	-	-	-	-
3		Car6f10	-	-	-	-	-	-	NP12-NN9
1		87-26a*	-	-	-	-	-	-	NP16-25
1		87-26b*	-	-	-	-	-	-	NP16-25
1		87-26c*	-	-	-	-	-	-	NP16-25
6		9-3-2	-	-	-	-	-	-	NP16-20
6		9-2-8p*	-	-	-	-	-	-	NP16-25
Belmont Formation B									
Fig. 15		87-2	-	-	-	-	-	-	NP16-NN17
Kendeace Formation									
Fig. 15		CarK2	-	-	-	-	-	-	NP12-NN9
Carriacou Formation									
Fig. 15		CarK1	-	-	-	-	-	-	NP25-NN8
Fig. 15		CarK3	-	-	-	-	-	-	-
Fig. 15		CarK5	-	-	-	-	-	-	NN4-5
Fig. 15		CarK8	-	-	-	-	-	-	-
Fig. 7		Carmp1	-	-	-	-	-	-	NN4-5
Fig. 7		Carmp2	-	-	-	-	-	-	NP25-now

example of thick units of the Anse La Roche Formation with only minor Quaternary cover. The section is perturbed mildly, however, by a complex of sills, plugs, and dikes of Neogene(?) basalt. Figure 13 shows a columnar section.

The lower 150 m (Fig. 13) is a succession of very thick (10 to 25 m) massive and graded units, which we interpret as debris and grain flows, depending on proportion of matrix. The debris flows are massive and matrix-bearing in the lower reaches but commonly are graded and stratified, well-sorted sandstone above. The grain flows are typically clast stratified and are thinner than debris flow units. The majority of gravel is intraclastic, and folded, twisted layering exists in some of the coarsest intraclasts. These thick units have scoured bottoms.

From 150 to about 250 m (Fig. 13), units are thick (3 to 17 m), graded grainflows with well-developed stratification and particle orientation. Coarse skeletal particles are as much as 25 to 75% of the particles coarser than sand in these units.

The section between 250 and 300 m is massive bouldery sandstone with unexposed base (Fig. 13).

As a whole, the Sparrow Bay succession may be interpreted as upward thinning/fining units in the lower 150 m and the inverse in the upper 150 m. If true, this implies the lower 150 m is a channel-filling sequence whereas the upper 150 m represents progradation over the filled channel, perhaps culminating in new channel cutting (now unexposed) below the top unit (bouldery sandstone at 250 to 300 m).

Dating. We dated hemipelagic beds at three sites and intraclasts at two sites in the Anse La Roche Formation (Figs. 9, 10, Tables 2, 3). Robinson and Jung (1972) and Westercamp and others (1985) also obtained microfossil ages of the Anse La Roche, each near Gun Point (Fig. 10). Our samples contain nannoflora and planktic foraminifers but no radiolarians.

Near Gun Point, the lowest exposed hemipelagic marl in the 3-m-thick section (Fig. 10, Site 1; Table 3) contains nannofossils that have a permissible range in zones NP17 to 21. We infer from the abundance of *D. deflanderi*, presence of *R. umbilica* and *E. formosa,* and absence of *D. saipanensis* and *D. barbadensis* that of this range, NP21 is most probable, hence an early early Oligocene age. Robinson and Jung (1972) identified planktic forams (Table 2) in the same outcrop (their 10764, Fig. 10) that give a permissible age range equivalent to nannozones NP17 to 20. Westercamp and others (1985) obtained planktic forams (Table 2) at unspecified sites in the Gun Point area that give a composite age range of NP19 to 20. Therefore, all identified species at sites near Gun Point overlap in NP19 to 20, late late Eocene. We think, however, that NP21 is the correct zone and

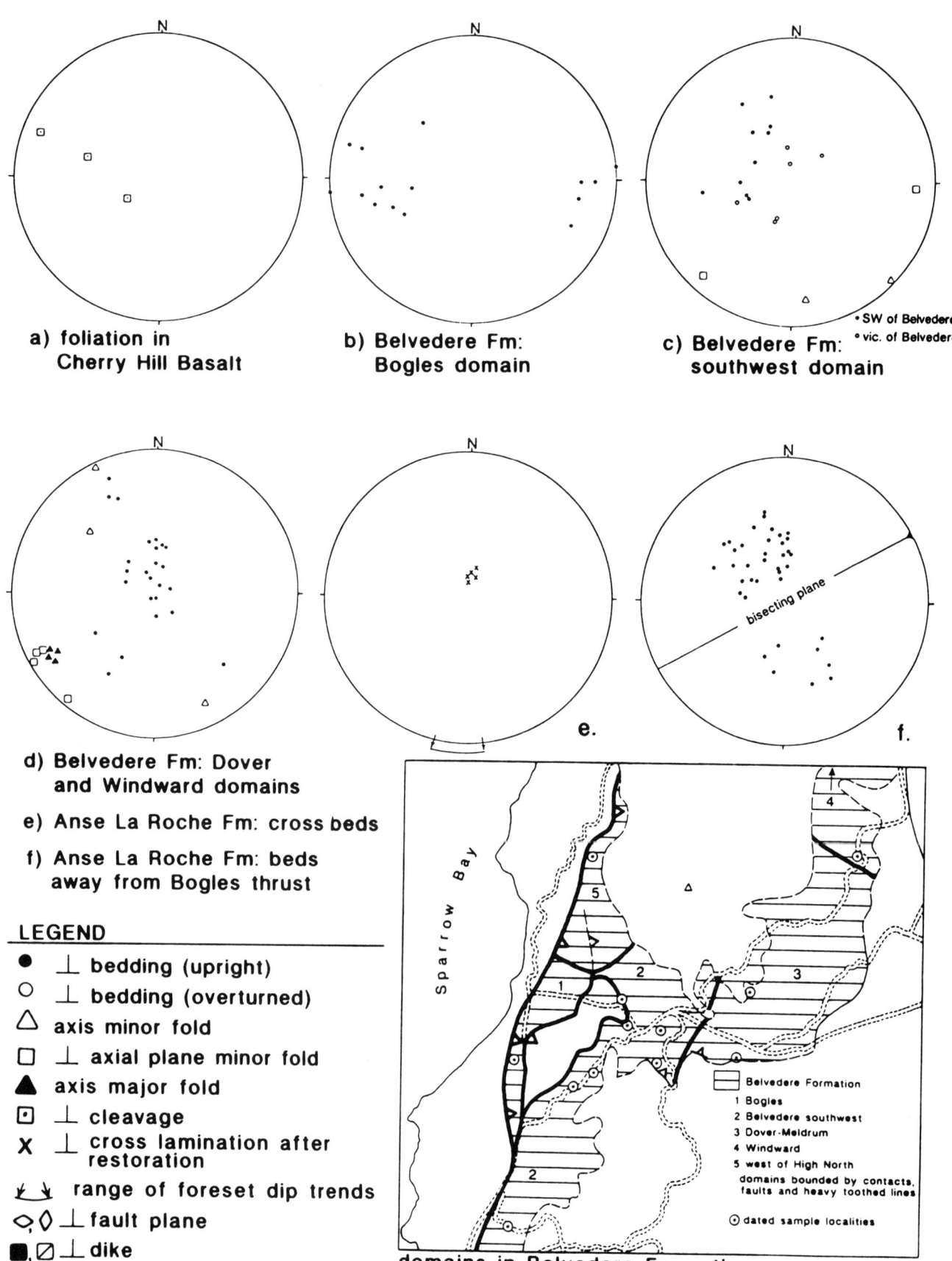

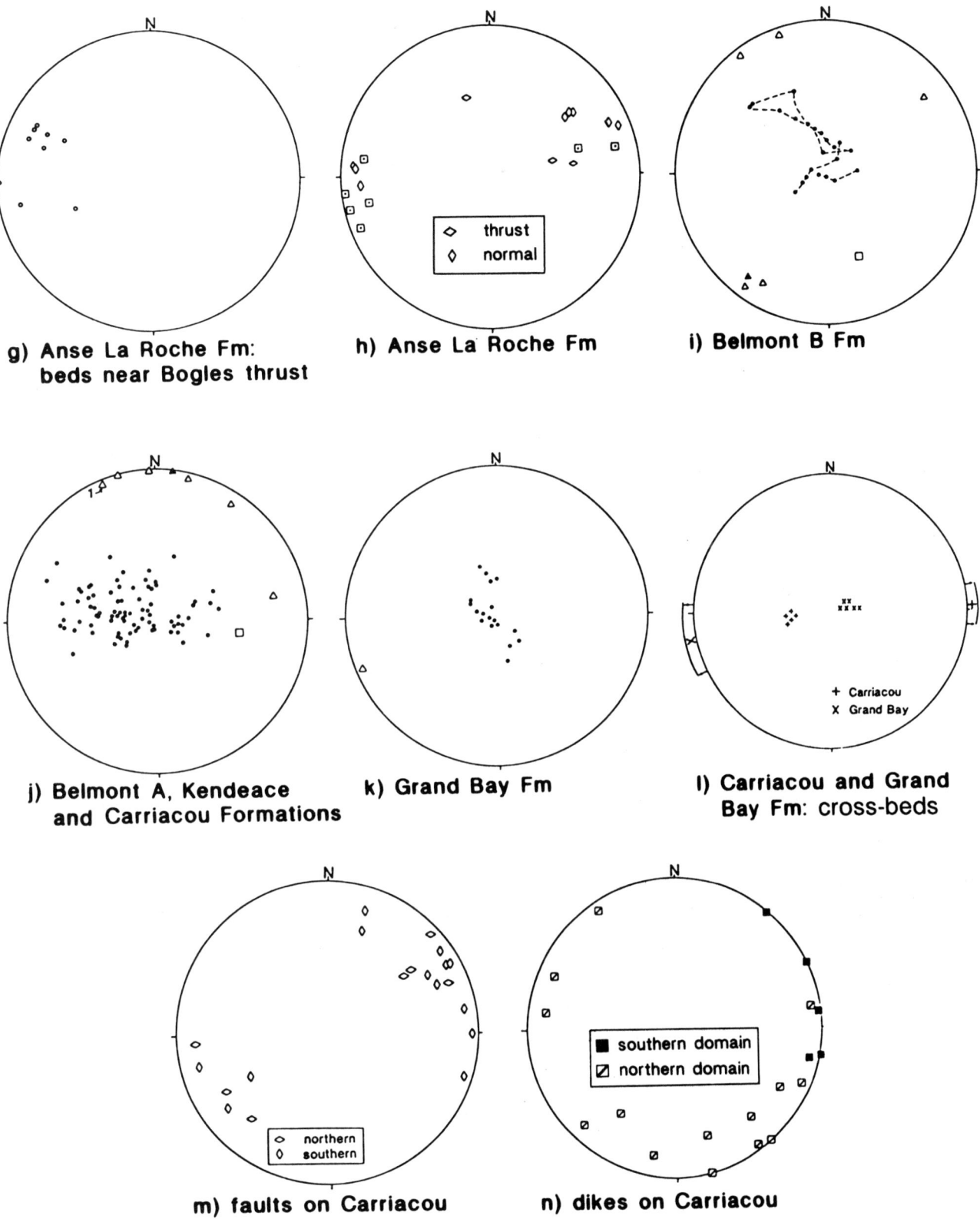

Figure 11 (on this and facing page). Map showing five domains of the Belvedere Formation and orientation diagrams for structural data of Carriacou; diagrams are equal-area nets, lower hemisphere; data also shown on maps of Figures 7, 10, and 15; measurements by P. L. Smith and R. Speed.

infer that certain planktic forams, identified by the other authors, notably *Hantkenina primitiva* and *Turborotalia cerroazulensis*, underwent minor resedimentation.

Our dated sample from a hemipelagic bed at northern Anse La Roche Bay (Fig. 10, site 2, Table 3) has a permissible range of NP17 to 20, late Eocene. Its content of *D. saipanensis* and *D. barbadensis* implies that the absence of these at site 1 is probably due to post-Eocene age at site 1 rather than preservation or different environment, as the lithotypes of sites 1 and 2 are very similar.

Our samples of hemipelagic beds, from southern Anse La Roche Bay (Fig. 10, site 3, Table 3) give permissible ranges no tighter than NP17 to 23, late Eocene to early Oligocene.

Our sample of a hemipelagic top of a grain flow in northern Sparrow Bay (Fig. 10, site 4) gave a permissible range of

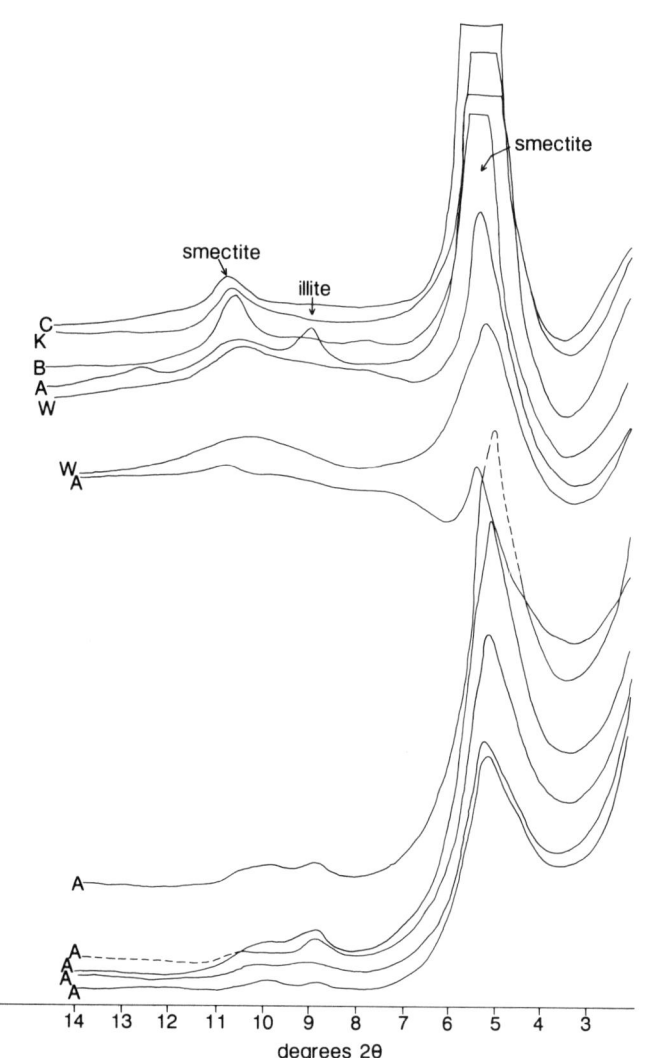

Figure 12. X-Ray diffraction patterns of clays in formations of Carriacou; CuKα radiation. A, Anse La Roche Formation; W, Belvedere Formation; B, Belmont B Formation; K, Kendeace Formation; C, Carriacou Formation.

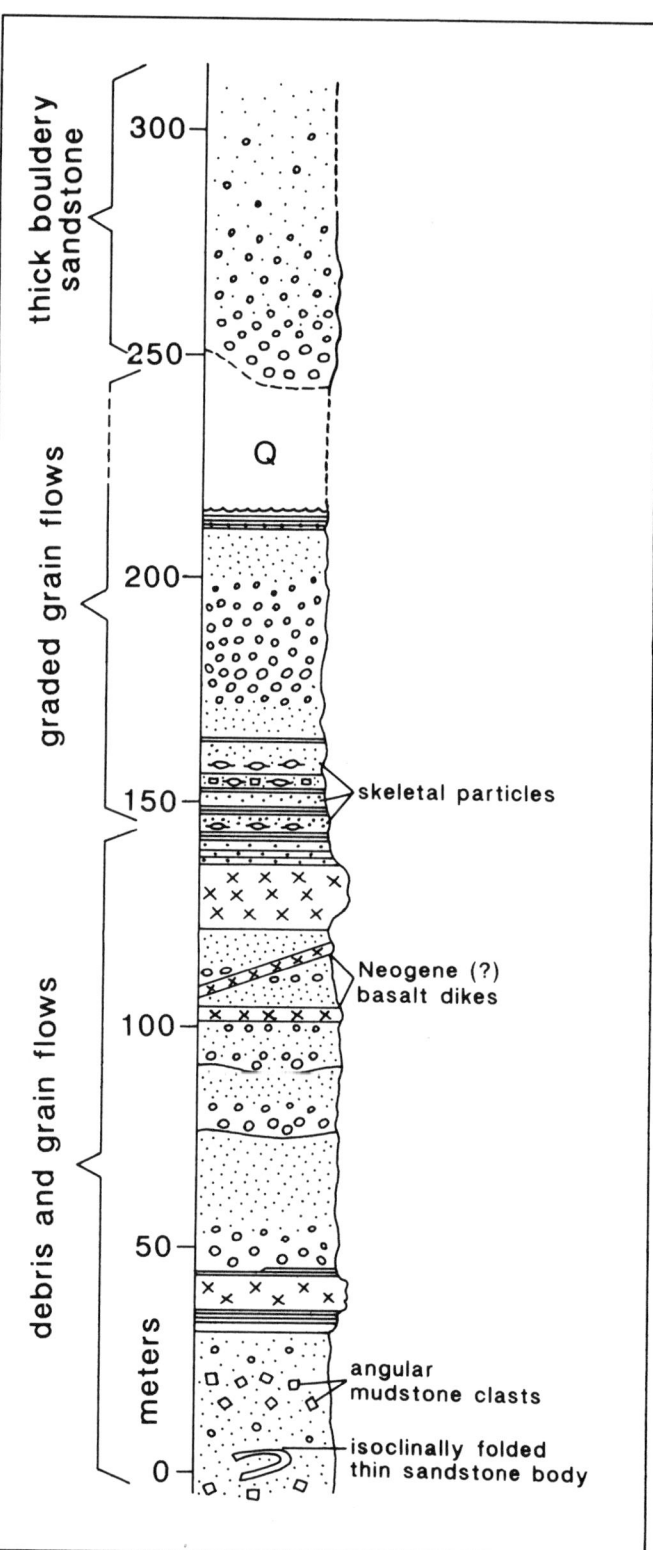

Figure 13. Stratigraphic section in Anse La Roche Formation at Sparrow Bay (Fig. 10) (measurements by P. L. Smith).

mid NP17 to NP19, based on *I. cocoaensis* and *C. reticulatum* (Tables 2, 3).

At site 1 near Gun Point, we also examined four marly intraclasts for nannoflora (Ca3, 87-26a, 26b, 26e; Fig. 10, site 1, Table 3). These yielded only long-ranging nanno species that gave permissible ranges of NP16 to 21. The source strata of the intraclasts are thus between late middle Eocene and early Oligocene age, assuming the beds at this site are lower Oligocene. The origin of the probably resedimented microfossils in the marl bed of site 1 may have been from partial disaggregation of such intraclasts. A marl intraclast in a pebbly sandstone near Prospect (Fig. 10, site 6, Table 3) gives a permissible range of mid-NP16 to NP20, late middle Eocene and late Eocene.

The larger benthic foraminifers of the Anse La Roche Formation are entirely resedimented. Species identified by Robinson and Jung (1972) and Westercamp and others (1985) are all thought to be approximately of Eocene age (E. Robinson, oral communication, 1986; Butterlin, 1981; J. P. Beckmann, written communication, 1986). Several species are thought to have first appearances in late middle Eocene time.

To conclude, zonal ages of the Anse La Roche Formation occur in the range NP17 to 21, late Eocene to early early Oligocene (Fig. 9). The unit's permissible range is only slightly greater, NP16 to 23. The Anse La Roche could include substantially older and (or) younger beds than either of these ranges because the stratigraphic range is very wide: unbounded lower limit to a early middle Miocene upper limit (Fig. 9). Resedimentation of some Eocene planktic forams probably occurred during Oligocene deposition. The duration between growth and resedimentation of larger benthic foraminifers in the Anse La Roche was short, less than about 1 to 7 m.y.

Sedimentology. Sediment gravity-flow deposits of the Anse La Roche Formation are in the main very thick bedded and coarse-grained grain and debris flows of volcanigenic and minor biogenic particles. Even with limited exposure, the bases of many can be seen to be scour or channel filling. Such features suggest accumulation in a channel system. The absence of intercalated pebbly mudstone implies that a slope is a less probable environment. The thin (0.1 to 1.0 m) sandy units are mainly tabular turbidites, which may have been intrachannel deposits during waning flows or overbank deposits. The hemipelagic component represents background sedimentation during relatively long intervals between sediment-gravity flows.

In terms of conceptual radial fan facies (Walker, 1984), the Anse La Roche contains mainly inner-fan channel and possibly, upper midfan deposits. Although such facies are considered proximal, there is little to gauge absolute distance to the source. The high proportion of grain flow (matrix poor) to debris flow (matrix rich) deposits among thick units implies that transport lengths were long enough for most flows to become watery and sorted.

The stratigraphy at northern Sparrow Bay indicates (Fig. 13) upward thinning and fining in the lower 150 m and upward thickening and coarsening in the upper 150 m. This suggests the filling of a major channel followed by progradation of a sandy fan-channel system (upper midfan and inner-fan channel facies) above the earlier channel.

The basin floor and (or) slope traversed by the Anse La Roche delivery system was at least partly underlain by hemipelagic sediments, represented by the marl and mudclasts that pervade the formation. It is a question whether such fragments are truly intraclastic, derived locally from semicohesive hemipelagic strata within the Anse La Roche Formation, or whether their source was upcurrent of the Anse La Roche basin. The middle or upper Eocene marl and mudstone of the Belvedere Formation are admissible in age and composition to have been the clast source.

The preservation of calcareous microfossils in hemipelagic beds of the Anse La Roche Formation indicates the basin floor was no deeper than 4.8 km in late Eocene time but permissibly as deep as 6.5 km in early Oligocene time.

The sources of allogenic debris in the Anse La Roche Formation were a subaerial volcanic terrane and a carbonate platform at neritic and (or) littoral depths. The volcanic terrane gave rise to all the siliciclastic particles, including the immature, nonmixed-layer smectite (Fig. 12). The volcanilithic gravel probably arose by sloughing of a beach rather than a pyroclastic flow down a slope channel because of the widespread rounding and mixture of volcanic lithotypes. Some thinner graded beds could have come from phreatic or primary eruptions, and some could be ash fall rather than turbidity current deposits. The granules composed of algal and foraminiferal debris are evidently from periodic sloughing from other sites, probably into the same channel system. The moderate mixing of carbonate and volcanigenic debris implies discrete source areas on the same platform. Thus, we postulate a volcanic island(s) partly rimmed by carbonate banks as the source region.

The volcanic source was mainly if not entirely of magmatic arc type, as indicated by the copious porphyritic lithotypes, including clinopyroxene and hornblende as phenocrysts. It is unclear whether aphyric or microporphyritic basalt clasts are arc derived or of a different kindred. The volcanigenic debris probably was derived from penecontemporaneous eruption rather than a then-ancient volcanic terrane. This is because none of the clasts is metamorphosed or tectonized.

Structure. Tectonic structures in the Anse La Roche Formation are the following, in the probable sequence of development: (1) train of macroscopic east-northeast–trending folds (Figs. 10, 11f), (2) local westerly overturned bedding below the Bogles thrust (Figs. 10, 11g), (3) minor thrust faults and related local cleavage (Fig. 11h), and (4) normal faults (Fig. 11h).

The train of close macroscopic folds is indicated by attitudes of locally homoclinal beds in coastal outcrops well away from or well below the Bogles thrust (Fig. 10). Bedding forms a broad girdle about a subhorizontal east-northeast–trending zonal axis (Fig. 11f), and facing control shows the folds are upright and have subvertical bisecting planes. Assuming layering is continuous along the coastal strip, bedding formlines (Fig. 10) illustrate the approximate positions of axial traces and show that wavelengths are 0.6 to 1.0 km, perhaps increasing northward. It is unclear whether the train dies out north of Anse La Roche or continues north with constant harmonic content. No foliation is associated with the east-northeast–trending fold train.

Local westerly overturning of bedding is evident near Bogles village (Figs. 10, 11g) but is unrecognized elsewhere in the Anse La Roche Formation. Exposures do not indicate the form of macroscopic structures in this area or the continuity of the overturned beds with upright beds in the east-northeast–trending folds of adjacent areas. We assume (sections, Fig. 10) that the overturned beds reflect superposed deformation on the east-northeast–trending folds in the immediate footwall of the Bogles thrust. Reasons are (1) the proximity of the overturning to the Bogles thrust and to the area in the Bogles thrust hangingwall of the major antiformal culmination that includes westerly overturning and overthrusting (Fig. 10), and (2) the evident discordance of hangingwall structures near Bogles to the apparently pervasive foldtrain of the Anse La Roche structurally down and west in the footwall. Therefore, we think that the hangingwall culmination and footwall overturning are related local heterogeneous strains that formed during or after emplacement of the Bogles thrust. We are unable to prove because of inadequate exposure, however, that the zone of overturning in the footwall is not a fault-bounded slice or that such overturning is not present everywhere in the immediate footwall of the Bogles thrust.

Thrust faults of small throw (<1 m) exist locally as singular or conjugate faults. Their dip is variable but chiefly east or west (Fig. 11h). Where thrusts cut outcrops of muddy or hemipelagic Anse La Roche beds, planar spaced cleavage exists with northerly strikes and steep dips (Figs. 11h). Such cleavage is later than the east-northeast–trending folds because it is discordant to their axial planes and exists at constant attitudes in beds of varied dips. Gaps in tilted calcareous laminae across some cleavage suggest dissolution and volume loss. The thrusts and associated cleavage indicate generally east-west contraction.

Normal faults of small displacement have northerly strikes as do a couple of Neogene(?) basalt dikes cutting the Anse La Roche Formation, each dated at about 10 Ma by Briden and others (1979). The dikes and normal faults cut the variably tilted bedding and appear uniformly post-tilt, and the normal faults in southern Anse La Roche Bay, at least, are younger than the cleavage. Therefore, the normal faults and dikes are probably the youngest structures (late Miocene or later) seen at outcrop scale. The attitudes of Neogene dikes in northwestern Carriacou, however, are highly varied, implying heterogeneity of strain over the whole of late Neogene time.

We now summarize the timing of the structures discussed above. The east-northeast–trending foldtrain developed either before or during the emplacement of the Bogles thrust and before the deposition of the unconformably covering Belmont strata. The westerly overturning near Bogles village was after the east-northeast–trending folding, either during or after the emplacement of the Bogles thrust. The north-striking thrusts and cleavage are younger than the east-northeast–trending fold train. By virtue of the similarity of their horizontal contraction direction with that affecting Miocene formations, they are probably mid-Miocene or younger. The normal faults are the youngest and are probably late Neogene.

Belmont Formation. The Belmont Formation (Figs. 7, 8) as defined here is composed of coarse volcanigenic sedimentary rocks that include only very minor proportions of carbonate clastics. Its base is an angular unconformity above the Belvedere and Anse La Roche Formations. Its top is an unconformity below the Kendeace and Carriacou Formations.

Stratigraphy and divisions. The stratigraphy and thickness of the Belmont are problematic, owing to a scarcity of age-indicative fossils and discontinuity of outcrop. Moreover, rocks called Belmont in southwestern Carriacou are as yet little studied. We divide rocks previously called Belmont (Belmont beds of Martin-Kaye, 1958; Belmont Formation of Robinson and Jung, 1972) into the Kendeace Formation, Belmont A, Belmont B, and Belmont undifferentiated; the latter encompasses the little-studied outcrop areas.

The Kendeace calcareous siltstone member was assigned to relatively calcareous strata at the top of the Belmont Formation by Robinson and Jung (1972). They designated no other members and did not map the Kendeace's distribution. We elevate the Kendeace to formational status because it is a mappable unit (Figs. 7, 8, 9, 10, 14, 15).

Belmont A comprises beds that lie unconformably below

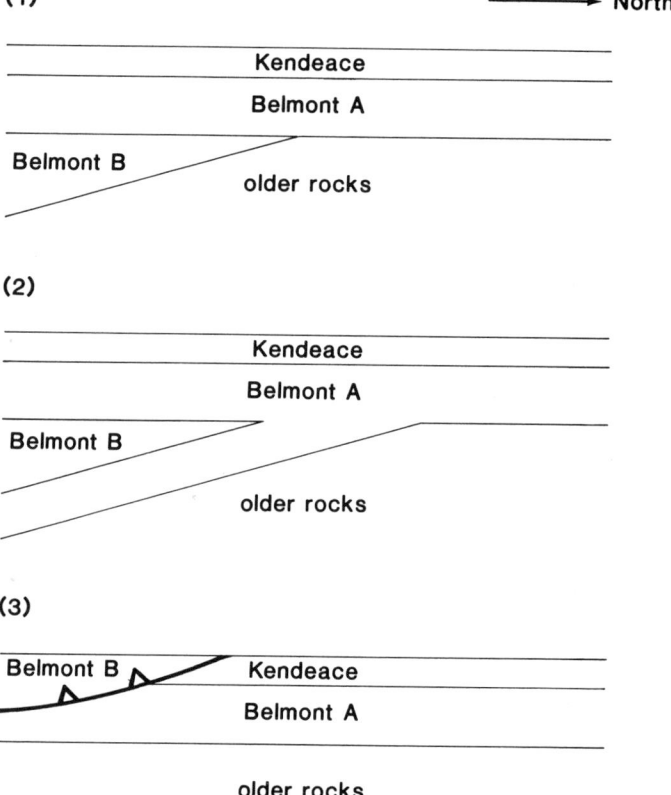

Figure 14. Alternative models for relationships of Belmont A and B beds.

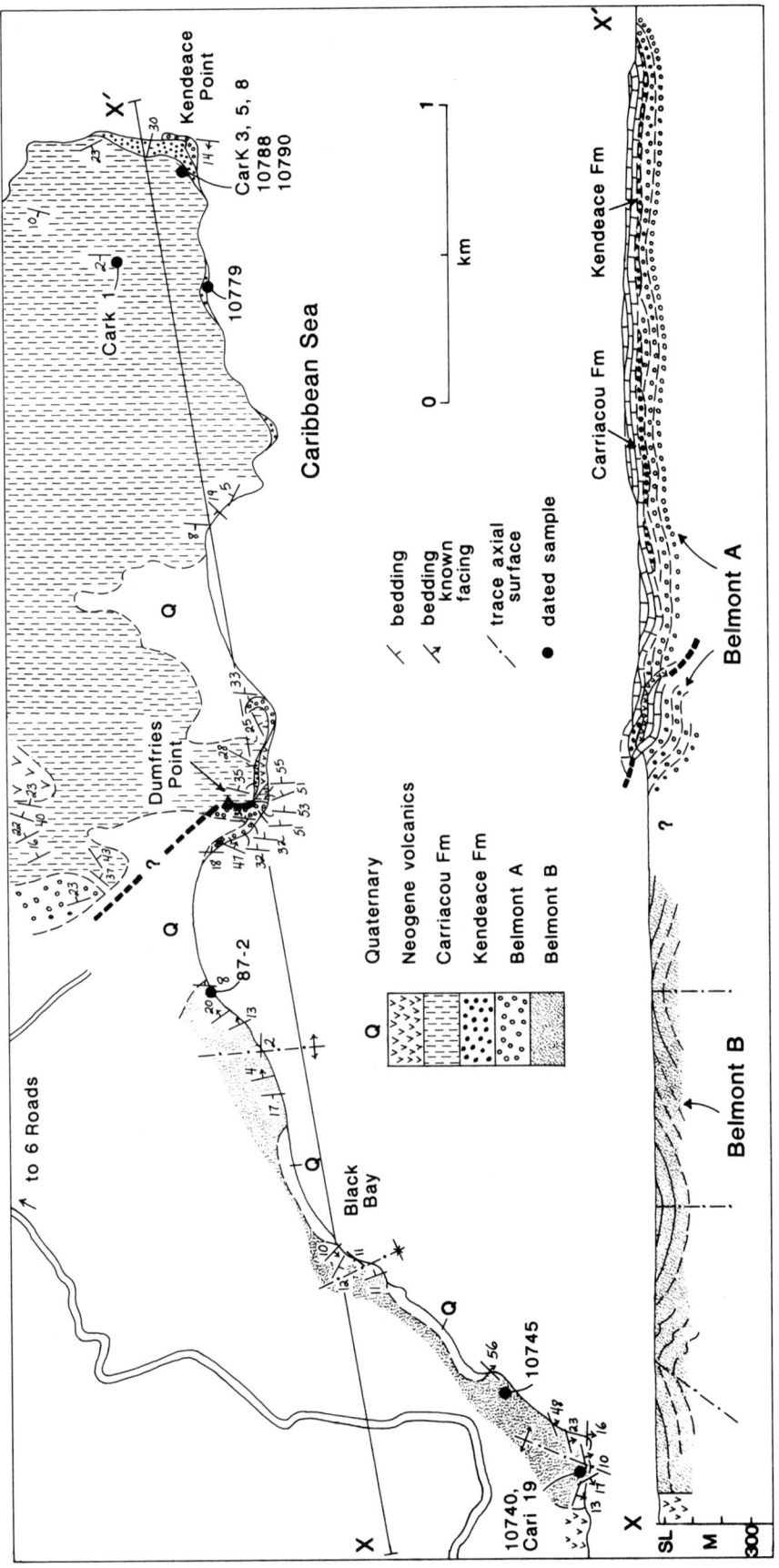

Figure 15. Geologic map and section of southern Carriacou; located on Figure 7; by R. Speed and P. L. Smith.

the Kendeace and Carriacou Formations in the east-tilted cuesta that forms the eastern slope of much of Carriacou. Belmont A is undated; its thickness varies between 100 and 200 m.

Belmont B is recognized only in southern Carriacou (Fig. 7). Because its outcrop area is circumscribed by Neogene rocks and Belmont undifferentiated, the contact relations of Belmont B with Miocene and older units are unknown. Belmont B, which is thicker than 275 m, contains the only dated beds (NN2 to 3, Fig. 9) in the formation, excluding the Kendeace. The predominant lithotypes of Belmont B differ in detail from those of A; if Belmont B extends north to central Carriacou in the stratigraphic interval between the Belvedere and Kendeace Formations, its distribution there is patchy. Therefore, the stratigraphic relations between Belmont A and B are in question. Possible alternative relations are as follows (Fig. 14): (1) Belmont B pinches out rapidly northward below Belmont A, (2) Belmont B pinches out rapidly northward as a facies within Belmont A, or (3) Belmont B is allochthonous and unrelated to Belmont A.

Belmont B. About 275 m of generally well bedded sedimentary rocks, here called Belmont B, crop out along part of the southern coast of Carriacou (Figs. 7, 15). Belmont B includes the only dated beds in the Belmont Formation (the former Kendeace member excluded) and those for which the Belmont Formation was originally defined (Martin-Kaye, 1958). Our interpretation of the structure of Belmont B (Fig. 15, explained below) indicates the succession is at least 275 m thick and could be far thicker. The Belmont B strata are sandy and gravelly sedimentary rocks of two interstratified compositional types: monolithic basaltic and polymict. Increasing proportions of polymict beds occur upsection. Each type contains massive, diorganized, clast-framework conglomerates that are chiefly channel-filling. They also both include graded sandstones that are either tabular or with strongly scoured bottoms, amalgamated zones, and sedimentary structures that define Ta, Tae, Tab, Tabe, Tabc, and Tabce zonations. Most beds are top-absent or are capped by a discontinuous mud drape.

Monolithic basaltic sedimentary breccia beds contain angular pebbles of crystal-rich (feldspar, olivine?, pyroxene), mainly vesicular basalt microporphyry. In contrast, the polymict beds contain angular and rounded cobbles and pebbles; the clast types include pyroxene porphyry, pyroxenite, feldsparite, hornblende porphyry, basalt, megafossil fragments including occasional articulated clams, and marlclasts. Within such beds, the proportion of fossil fragments increases upsection, although it is always subordinate. A few thin discrete tabular beds of microfossilierous smectitic mudstone exist (Fig. 12), evidently recording pauses in the influx of coarser debris.

We interpret the beds of Belmont B to be subwavebase, probably deepwater deposits from the following evidence: (a) the turbiditic features (tabularity, upward fining, and upward diminishing flow features) of most sandy beds, (b) intercalation of discrete muddy layers with open-marine planktic faunas, (c) the evident resedimentation of neritic macrofossils, and (d) inclusion of poorly lithified marlclasts in conglomerates, implying sediment transport through and wall collapse of open sea channels.

Beds in the Belmont undifferentiated unit that lithically resemble Belmont B occur at Sugar Hill and at places on Chapeau Carré (Fig. 7).

We found no nannofossils or radiolarians in several specimens of mudstone beds of Belmont B. Zonal age ranges of N5 to 6 for beds apparently low and high in the Belmont B section were obtained, respectively, by Robinson and Jung (1972, their 10745) and Westercamp and others (1985, their Caril9; Figs. 9, 15, Table 2). These indicate Belmont B is at least mainly early Miocene. Nannofossils in a marlclast in the lower section are wide-ranging but not younger than NN7 (middle Miocene; Table 3). A K-Ar whole-rock date of 18.1 ± 0.5 Ma was measured from a clast of basalt in Belmont B by Briden and others (1979; Fig. 9). This implies that eruption of basalt at the sediment source was nearly contemporaneous with Belmont B deposition. However, it cannot be inferred that the nonbasaltic rocks of the polymict beds were also coeval.

Rocks of Belmont B had two evident sediment sources. One was almost exclusively of fragments of contemporaneous basalt and was probably submarine as suggested by the absence of particle abrasion. The second included mixed terrains: subaerial gravels, probably on the lower flanks of an arc-type volcano, and shallow-marine areas. As in the Anse La Roche Formation, we interpret the lack of metamorphism, alteration, and deformation of the arc-derived clasts to indicate penecontemporaneous volcanism. We envision channelized sediment transport from an arc platform through a zone of basaltic magmatism on or below the slope to a lower slope or basin plain site. The occurrences of marlclasts in the sedimentary breccias and of interbeds of mudstone bearing planktic forams support the interpretation.

Belmont A. Belmont A beds are mainly polymict volcanigenic boulder sedimentary breccia and conglomerate together with nonpebbly sandstone. The thickness is between 100 and 200 m.

Belmont A beds occupy the western rim of the extensive east-dipping cuesta capped by Carriacou Formation (Fig. 7). Belmont A at the west side of Dumfries Point (Figs. 7, 15) contains well-exposed typical lithotypes. The basal sediments at Kendeace Point (Figs. 7, 15) may be part of the same section. Similar coarse volcanigenic sediments with the same northerly strike and gentle easterly dip make up most of the central ridge of Carriacou, from Dumfries through Mt. Desire to Belvedere, and both flanks of High North (Fig. 7). Macrofossil fragments and marlclasts occur in some of these coarse rocks, as do rare mudstone interbeds bearing larger benthic forams (Robinson and Jung, 1972). The beds at Dumfries Point are occasionally graded but are not classic turbidites, whereas the section from Mount Desire to the Forest Reserve and on the west flank of High North contains at least a few meters of sandy and muddy turbidites with Tabe and Tbce sequences. We interpret the coarser rocks as submarine debris flows and channel-filling sequences, perhaps in a proximal-fan setting where occasional, finer turbidites were deposited.

There are no fossil dates of Belmont A. Its minimum stratigraphic age is NN5, early middle Miocene, the minimum fossil age of the unconformably overlying Kendeace Formation (Fig. 9). If Belmont A beds overlic Belmont B (Fig. 14, model 1), the maximum stratigraphic age of Belmont A is NN3, late early Miocene. If, however, Belmont B is allochthonous, or a facies within A, the maximum stratigraphic age of Belmont A is NP25, late Oligocene, the youngest age of the unconformably underlying Belvedere Formation (Fig. 14, model 3).

Beds of Belmont A are sediment-gravity flow deposits that accumulated in subwavebase conditions. The prevalence of gravel flows in Belmont A and the at-least partial lacuna at the top of the Belvedere Formation suggest Belmont may fill a major channel system. The sandrich intervals may be overbank and (or) waning flow deposits. The sediment was from a magmatic arc source. Unlike Belmont B, Belmont A had no strictly basaltic source at any time in its depositional history.

Belmont structures. Strata of Belmont B are in macroscopic open folds, which may occupy a continuous wavetrain (Fig. 15). Each fold is approximately of 2 km wavelength. The axial traces trend NNW to NNE but axial plane orientations are poorly controlled. Zonal axes plunge shallowly southwest and north-northwest (Fig. 11i). There is no associated foliation. The western anticline of the wavetrain may have an inclined axial plane, dipping moderately northwest (Fig. 15). This is implied by the relatively steep bedding and the existence of two small folds on the fold's eastern limb (Fig. 15). The small folds are homoaxial with the anticline, implying a harmonic set. The axial plane of a minor fold is northwest-dipping, suggesting that of the major fold may be similarly inclined.

Belmont A is generally poorly stratified and contains no evident microstructures, such that it is difficult to determine its inherent structures. Where bedding has been recognized, however, its attitude is similar to that in the overlying Kendeace and Carriacou Formations, and we employ the grouped bedding distribution, as indicative of structures for all three formations (Figs. 10, 11, 15). The distribution includes a gently east dipping homocline, the cuesta of western Carriacou, and a broad girdle with a horizontal, north-trending axis (Fig. 11j). A half-dozen singular folds have been identified in the Belmont A contact and succeeding Miocene strata. These are 50 to 150 m wide and open. The axes of five such folds are subhorizontal and NNW to NNE-trending whereas the sixth plunges moderately eastward (Fig. 11j). The variability of axial trends of such folds explains why the girdle of Figure 11j is so broad.

Fold 1 in Belmont A and the Kendeace Formation (Figs. 10, 11j) occurs 200 m south of Belvedere. The fold may record deformation in the footwall of the nearby northwest-striking reverse fault that places Belvedere Formation above Belmont A. Fold 1 and the reverse fault are both probably part of a broader deformation that caused refolding of the macroscopic anticline in the Belvedore Formation in post-Belmont time (Fig. 10).

The fold of Belmont A and succeeding units at Dumfries Point (Fig. 15) is in the footwall of a thrust fault of probably minor displacement. A thrust ramp lies just east of the fold. The thrust zone is occupied by a Neogene intrusion and by tabular diatremes of metamorphosed fragments of sedimentary rocks, chilled igneous fragments, and tuff. It is unclear whether thrusting and intrusion were concurrent or sequential.

Kendeace Formation. The Kendeace Formation contains moderately to well-bedded clastic sediments of mixed volcanigenic and carbonate compositions. It is conformable and gradational with the suprajacent carbonaterich Carriacou Formation and locally unconformable above the generally massive carbonate-poor A unit of the Belmont Formation. As noted above, the here-named Kendeace Formation was previously called the Kendeace calcareous siltstone member of the Belmont Formation. We have changed its status because of its mappability, and moreover, to call attention to its record of transition in sediment provenance and depositional conditions between the Belmont and Carriacou Formations.

Rocks. The Kendeace is preponderantly sandstone and fine pebbly sandstone; arenite and wacke are both prevalent. Conglomerate and sandy mudstone also occur. The strata are commonly distinctively bedded owing to rapid vertical changes in grain size, including both abrupt and gradational transitions, and in proportions of volcanigenic and carbonate sand and gravel. Lateral stratigraphic changes are at least partly related to channels filled by Kendeace in the subjacent Belmont Formation.

Volcanigenic grains in sandstone are feldspar, dark minerals, lithic fragments, and carbonate grains are skeletal fragments, pellets, and algal material. The ratio of volcanigenic to carbonate grains takes on a broad range. It varies markedly bed by bed in some successions, but in general, decreases upward through the formation. The base of the Carriacou Formation is arbitrarily placed below the lowest ledgy calcarenite.

Grains coarser than sand in the Kendeace are volcanilithic, mainly fine-grained porphyries, and skeletal and algal types: foraminifers, oysters, gastropods, algal balls, pteropods, and other types of megafossils (Robinson and Jung, 1972). Some large bivalves, including oysters, are whole and little abraded and are possibly infaunal. Wacke, fine-grained arenite, and mudstone in the Kendeace is consistently calcareous, presumably due to lime-mud particles in the matrix.

Thickness and channelization. The Kendeace is discontinuous above the Belmont (Figs. 7, 15) and varies in thickness from 0 to 40 m. The variability reflects in part deposition of the Kendeace on an eroded Belmont surface. The unconformity is best exposed at Kendeace Point (Figs. 7, 15), where the Kendeace fills a channel of amplitude >10 m and width >100 m. The exposed southern wall of the channel has a stairstep configuration with vertical risers cutting across the subhorizontal noncalcareous beds of the Belmont Formation. The channel margin facies of the Kendeace consists of amalgamated lenses of graded scour-filling sandstones and pebbly sandstone, each with abundant skeletal grains, and locally skeletal hash. The channel trough facies includes more tabular and somewhat more carbonaterich but less pebbly sandstones. The sandstones are generally medium bedded, and some are muddy and (or) fine upward. Most have no Bouma zonations. Bioturbation is indicated by vertical and horizontal tubules.

Above the channel fill at Kendeace Point, the Kendeace Formation contains about 13 m of moderately bedded sediments that are interfingering massive wacke; sandy mudstone; medium- to coarse-grained, plane-laminated sandstone; and occasional lensy pebbly to boulder sandstone. The coarsest beds contain large oysters and mud intraclasts together with volcanic clasts. Calcareous components exist as fine matrix as well as sand and coarser grains.

Dating. Our search for nannoflora at Kendeace Point gave meager results (Table 3). Robinson and Jung (1972) obtained planktic foraminifera from samples of Kendeace at six sites (Table 2; Figs. 7, 10, 15). None gives a zonal age, but two, 10788 and 10790 from Kendeace Point (Fig. 15), have short permissible ranges of mid-N6 to N8, late early to early middle Miocene (Fig. 9). The stratigraphic age range of the Kendeace is from NP25, the zonal age of the highest dated beds in the Belvedere Formation, and NN6, the top of the oldest zonal age of the overlying Carriacou Formation (Fig. 9). The stratigraphic range thus permits limits of late late Oligocene to middle middle Miocene for the Kendeace.

Depositional environment. The Kendeace Formation records a transition in clastic particle provenance and probably in conditions at the site of deposition between the Belmont and Carriacou Formations, following an episode of erosion of the Belmont A beds. We recognize no deformation associated with pre-Kendeace erosion.

A shallowing of the Kendeace site relative to that of the Belmont is suggested by the limey mud matrix, diversity of megafossils, possible infauna, the pteropods, which are aragonitic, and copious bioturbation in the Kendeace beds. Further, the increased influx of littoral and neritic carbonate debris during Kendeace relative to Belmont time, heralding the Carriacou Formation, suggests a migration of the site toward a carbonate platform and (or) shallowing toward platformal depths during Kendeace time. There is no evident trend of shallowing in the Belmont Formation leading to the Kendeace, but if such strata existed in the Belmont, they may have been eroded before Kendeace deposition.

We speculate that the unconformity below the Kendeace was caused by uplift and erosion of the basinal region containing the Belmont deposits. After the site shallowed, possibly to the neritic zone, and attained a new configuration, the Kendeace was deposited as a shallow-marine clastic complex, mainly sandy. The Kendeace sediment was delivered from one or more arc volcanoes and from a carbonate platform that prograded across the Kendeace site in Carriacou Formation time (middle Miocene).

Structure. The structure of the Kendeace Formation is like that of the Belmont A beds and the Carriacou Formation and is discussed with those units.

Carriacou Formation. *Rocks.* This unit, 80 to 100 m thick, is distinguished from other Miocene rocks of Carriacou by its high proportion of carbonate, including in some sections up to 25 m of nearly pure limestone. Calcarenites form the bulk of the Carriacou Formation. These are medium to very coarse grained, thin to thick bedded, and composed mainly of skeletal particles, together with minor volcanigenic grains. Some beds are well-cemented, algal-large benthic foram deposits that include large mollusc fragments, whereas in other sections soft calcwacke predominates. Bedding is defined by grain-size differences, but only a few beds are clearly graded. Some sequences are composed of thick cross-laminated beds. At Point St. Hilaire (Fig. 7), a stack of roughly 80 beds has uniformly east dipping foresets after correction for tilt by rotation to horizontal about the strike of bedding surfaces between sets of foresets (Fig. 11l). Each foreset set is between 0.5 and 1 m amplitude. Smectite mudstone (Fig. 12) occurs sparingly in the Carriacou as fossiliferous thin beds intercalated with but sharply contacting calcarenites. As noted by Robinson and Jung (1972, Fig. 4), lithofacies of the Carriacou are markedly variable, and no two sections are identical.

Dating. We found useful nannofossils in two samples of mudstone of the Carriacou Formation; these indicate zonal ages of NN4 to 5 (Fig. 9, Table 3). These are CarK5 at Kendeace Point (Fig. 15) and CarMp1, which is 2 km north (Fig. 7). Robinson and Jung (1972) established a type section on the southern side of Kendeace Point and there divided the formation into three units. They obtained planktic foraminfers from the lowest and highest of these (Table 2). The lowest, up to 10 m thick, gives a composite age range of N7 to 8, and the highest, up to 25 m thick, has a composite age of N8 (Fig. 9). The stratigraphic range of the Carriacou Formation is N6 to 10. The foram ages indicate that the bulk of the formation is early middle Miocene, but the base and top could be slightly older and younger, respectively (Fig. 9).

Deposition. The Carriacou Formation contains a spectrum of carbonate depositional environments that were generally remote but not excluded from influx of volcanigenic sediments. All of the sand-sized and coarser particles are allogenic except for infauna, but at only a few places are there graded beds suggestive of turbidity current transport. The muddy, planktic foram-bearing beds and soft, muddy calcwacke are suggestive of subtidal or lagoonal accumulations. The thick successions of cross-laminated beds, however, are more likely intra or supratidal. The Carriacou Formation thus is probably a complex of platformal and periplatformal deposits.

Cross-laminations in the Carriacou Formation at Point St. Hilaire (Fig. 7) imply easterly sediment transport in today's coordinates (Fig. 11l), perhaps reflecting an east-facing Miocene beach.

Structure. The Carriacou Formation, together with underlying beds of Belmont A and Kendeace Formations and overlying Grand Bay Formation, contain open folds of 10- to 100-m widths with mainly north-trending axes (Fig. 11j). This stratigraphic succession occupies an east- to east-southeast–dipping slope on eastern Carriacou (Fig. 7) that is an enveloping surface to the folds. The dipslope may be the eastern flank of an arch whose axial region is underlain mainly by Neogene magmatic rocks.

The Carriacou Formation also includes a fold with an east-plunging axis (Fig. 11j) at Point St. Hilaire (Fig. 7). This fold occurs in the wall of a 30-m-wide diatreme that cuts vertically through the Carriacou Formation and contains rounded volcanic cobbles and sand and limestone blocks. The fold is presumably related to diatreme emplacement.

Grand Bay Formation. The Grand Bay Formation (Robinson and Jung, 1972) contains volcanigenic sedimentary rocks that lie above the Carriacou Formation. The Grand Bay is the highest Miocene sedimentary unit. The rocks are volcanigenic sandstone in thick, commonly fully cross-laminated beds, channel-filling volcanic conglomerate and pebbly sandstone, and a few debris flows. Megafossils occur commonly. At places, the Grand Bay contains microfossiliferous marly interbeds. Planktic foraminifers identified by Robinson and Jung (1972) in Grand Bay specimens give seven zonal ages that include ranges of N8 to 10 and 10 to 11 (Fig. 9). The stratigraphic range is N8 to 11 Ma (Fig. 9). The Grand Bay Formation is thus entirely middle Miocene, including middle middle Miocene.

Grand Bay strata point to flooding of the varied carbonate depositional sites of the Carriacou Formation by volcanigenic debris in middle Miocene time. At least the bulk of the formation is marine, but indicators of its bathymetry are lacking. The prevalence of sandstone with steep dunelike cross-laminations of high amplitude suggests it partly originated as beach deposits. In contrast to the eastward progradation of cross-laminations in the Carriacou Formation, the cross-laminae from the Grand Bay indicate west to southwest progradation (Fig. 11l).

Measured bedding in the Grand Bay is shallowly dipping and may lie on the same girdle as bedding of the Carriacou Formation (Fig. 11k). One open fold was seen in the Grand Bay; its axis trends east-northeast, quite askew to the girdle in other Miocene formations (Fig. 11j). The fold is faulted, however, and it may be of fault-accommodation or diatremal origin and unrelated to regional deformation.

Petit Martinique. Petit Martinique is a small island of 2 km^2 area about 4 km east of northeastern Carriacou (Fig. 1b). The island is mainly underlain by a layered succession of intercalated arc-volcanigenic and carbonate sediments (Martin-Kaye, 1969; Westercamp and others, 1985). Calcareous beds within the succession were dated by many species of planktic foraminifera with concordant ranges in the *Globorotalia fohsi lobata* zone, upper N11 and N12, middle Miocene, approximately 12.8 to 13.2 Ma, by J. B. Saunders and by Westercamp and others (1985). The dated beds on Petit Martinique are thus contemporaneous with or slightly younger than the Grand Bay Formation of Carriacou. Like the Grand Bay, they are richly cross-bedded, implying a similar platformal environment of deposition. Moreover, their principal dip is easterly, as in the Kendeace-Grand Bay sequence of Carriacou. Dikes and a plug of basalt intrude the Miocene beds of Petit Martinique.

Young structures. High-angle faults and dikes cut all the layered rocks discussed above. Their orientations are shown in Figure 11m and n, divided into two domains by area and age of rocks cut: (1) a northern domain of the Paleogene Belvedere and Anse La Roche Formations, and (2) southern domain of the Miocene formations.

The faults are mainly normal; only two of the population shown are reverse. Strikes are mainly between north and northwest (Fig. 11m). The faults indicate horizontal extension between east-west and northeast to southwest trends affected Carriacou at one or more times since the middle Miocene.

The tabular dikes shown in Figure 11n are steeply dipping. The strikes of dikes in the southern domain are like those of islandwide normal faults whereas those in the northern domain have no evident concentration. Further, coastal exposures in northwestern Carriacou of the Anse La Roche Formation include local nests of irregular intrusions: sills, transitional sills and dikes, curvy dikes, splayed dikelets at major dike tips, and plugs; most of these are not plotted on Figure 11n. Therefore, dikes at structurally high levels of Carriacou (south) record a generally similar horizontal extension orientation as islandwide normal faults. The lower level of intrusion in the north, however, does not provide a systematic kinematic record. The ages of intrusions on Carriacou are not fully known, but the dates given by Briden and others (1979) are not indicative of different intrusive episodes in the two areas.

The geologic map of Carriacou (Fig. 7) shows that outcrops of older premagmatic units and Neogene igneous rocks are concentrated in western Carriacou. In contrast, Miocene sedimentary rocks underlie most of eastern Carriacou where they form a shallowly east dipping cuesta that contains a minor quantity of intrusion except at High North. This arrangement permits several alternative interpretations: (1) the locus of late Neogene magmatism is in the crestal zone of a north-south–trending arch that is in or just west of western Carriacou; (2) Neogene magmas mainly rose to a level below or within lower Miocene (Belmont) strata before easterly tilting of the cuesta; or (3) western Carriacou is a zone of large extension, unroofing, and intrusion; higher strata have detached and moved east and perhaps west of this zone. The second hypothesis predicts west-dipping rotated dikes in the cuesta. The few dikes observed in the cuesta are vertical, suggesting this idea is unlikely. The third hypothesis predicts substantial extension in western Carriacou that is not in harmony with our observations. Further, it implies a flat fault exists at the base of or within the Miocene succession. We have not recognized such a fault. Its existence is conceivable at the base of the Belmont A beds, but if so, no evident fault-zone structures were generated. To conclude, the locus of magmatism in the crest of an arch seems the most plausible solution.

Discussion: Depositional history of rocks of Carriacou. There are four major tectonostratigraphic units of Carriacou (Fig. 8): (1) the autochthon to the Bogles thrust, the Anse La Roche Formation; (2) the Bogles allochthon, comprising the Cherry Hill Basalt, Bogles Limestone, and Belvedere Formations; (3) Miocene sedimentary cover on the Bogles thrust; and (4) Neogene magmatic rocks (11 Ma and younger).

The Anse La Roche Formation records late Eocene–early Oligocene and perhaps a greater duration of basinal environments to which sediment-gravity flows delivered volcanigenic and skeletal debris from one or more shoaled arc volcanoes and marginal platforms.

The formations of the Bogles allochthon indicate middle Eocene extrusion of pillow lava in a basin receiving pelagic carbonate deposition (Bogles Limestone), followed by a long duration, middle Eocene to late Oligocene or possibly, early Miocene, of basinal deposition of arc-derived turbidite and pelagic carbonate (Belvedere Formation). If our correlation of the Cherry Hill Basalt with Mayreau Basalt is correct, the middle Eocene basalt was a product of spreading, not of arc volcanism (Speed and Walker, 1991), and the oceanic basin later became the repository for arc-derived sediment, as first recorded on Carriacou in the Belvedere Formation.

The Anse La Roche and Belvedere Formations evidently overlap in time, at least in late Eocene and early Oligocene. Moreover, they are alike in types of constituent particles and in basinal deposition. They differ greatly, however, in grain size range and proportions of particle types and in flow regimes during deposition. The Anse La Roche is generally high regime and channel-related whereas the Belvedere is a mainly low regime lower fan and hemipelagic deposit. An important question is the distance between the depositional sites of these two formations, or equivalently, the displacement magnitude of the Bogles thrust. We think the two formations are not local facies because of the paucity of interfingering in either formation of rocks characteristic of the other and because the time overlap may be substantial, about 3 m.y. or more. On the other hand, such differences imply the two formations may initially have been far apart (tens of kilometers). It is tempting to suggest that the marly intraclasts in the Anse La Roche were obtained during transit through the Belvedere by channelized sediment-gravity flows. We believe, however, that the provenance of the intraclasts is too nonunique to press this idea forward. To conclude, we are uncertain of the late Eocene–early Oligocene paleogeography of the Anse La Roche and Belvedere Formations.

The Belmont Formation contains sediment-gravity flows that were derived mainly from shoaled-arc volcanoes and in the Belmont B beds, from a probably nearby, isolated basalt vent. The Belmont sediments were laid down in channels and fans on the eroded surface of the deformed Bogles allochthon and autochthon at depths below wavebase. Belmont deposition certainly occurred in early Miocene time and may have extended back into late Oligocene time. The volcanoes at the Belmont provenance were part of the Neogene chain, as indicated by overlapping ages of Neogene magmatic rocks in the SLAAP as a whole and Belmont deposition. Carriacou, therefore, was still a basin in the early Miocene whereas some other regions had emerged as magmatic platforms.

The Kendeace Formation of late early and (or) early middle Miocene age covers the channeled, unconformable surface of the Belmont Formation. The Kendeace records the increasing influence upsection of a carbonate platform, which by middle Miocene time, prograded over Carriacou as the Carriacou Formation. Erosion recorded by the unconformity between the Belmont and Kendeace Formations can be inferred to have accompanied uplift from basinal (Belmont) to periplatformal (Kendeace) conditions. Carriacou, therefore, shoaled at least 1 to 7 m.y. after the regions that were sources of sediment to the Belmont Formation (Fig. 9).

The Grand Bay Formation records progradation of volcanigenic sediment from easterly sources above the carbonate platform of Carriacou during part of middle Miocene time.

Development of major structures. The principal structural events of Carriacou were (1) emplacement of the Bogles thrust; (2) early folding of the Anse La Roche beds on east-northeast–trending axial traces; (3) development of the macroscopic anticline and its apparent culmination in the Bogles allochthon; and (4) folding on northerly axes. Event 1 occurred at a time within the duration from late Oligocene to early middle Miocene, between about 15 and 28 Ma, as indicated by youngest strata of the Belvedere Formation cut by the thrust and by the minimum age bound for the overlapping Belmont A Formation (Fig. 9). The emplacement was before about 19 Ma if Belmont B was deposited in place. Event 2 occurred before event 1 but not before early Oligocene time. Event 3 may have been synchronous with or after event 1, probably both, and before and after Belmont A deposition (section Y–Y′, Fig. 10, and discussion below). Event 4 is post–Carriacou Formation, hence between middle Miocene, about 17 Ma, and the earliest Neogene magmatism, 11 Ma.

Bogles thrust. The Bogles thrust is clearly the most important tectonic feature of Carriacou because it juxtaposes rocks that are contemporaneous over at least several millions of years and that are of substantially different lithofacies. Therefore, the thrust is of more than local extent, and a heave of tens of kilometers is conceivable. There are no direct indicators of displacement direction, although we infer below that the trend of slip was approximately northwest-southeast for at least part of the thrust's movement history. Features in both walls of the Bogles thrust indicate brittle behavior and a probably shallow and low-temperature deformation environment at all stages of motion. The dip of the Bogles thrust is easterly (Fig. 10) and interpreted to vary from subhorizontal to 45°.

Anse La Roche folds. The early structures of the Anse La Roche Formation appear to be a train of upright folds with subvertical axial planes except at the site of anomalous noncylindrical westerly overturning near Bogles village and possibly, more widely in a 20- to 40-m-thick zone below the Bogles thrust. The early folds formed in the brittle realm, evidently by flexural slip without associated flattening. Our interpretation that the early Anse La Roche folds are in a train implies an origin by buckling rather than by accommodation to fault bends, and the existence of opposing limb dips of about the same magnitude implies the folds are not rollovers in listric normal fault hangingwalls. The fold geometry indicates subhorizontal principal contraction approximately NNW-SSE.

At question is whether the early folding of the Anse La Roche beds preceded or is related to the emplacement of the Bogles allochthon. A relation to thrusting is possible because the trends of axial traces of major structures in both walls of the Bogles thrust may have been similar. If the second-phase (post early Miocene) folding of the hangingwall symmetrically rotated the early macroscopic anticline's axial trace from a straight line, the initial trend of the axial trace could have been like that of the folds of the Anse La Roche beds. Therefore, both hanging- and footwall folds may have developed during the same plane strain with a north-northwest–striking, subvertical XZ plane. Because the early Anse La Roche folds appear to be a train and because there is no evident thrust-parallel simple shear in the foldtrain, however, the footwall folding probably did not occur below the Bogles thrust. Rather, the foldtrain was more likely a product of layer-parallel shortening ahead of the encroaching Bogles allochthon. The folded Anse La Roche was then overrun by the Bogles allochthon.

Bogles allochthon. The dominant anticline of the Bogles allochthon is inclined, almost overturned, to the northwest and is internally thrust at its culmination near Bogles. There is, however, no evidence for significant inclination or internal thrusting of the dominant anticline east or south of the culmination where the surface intersects much higher stratigraphic levels. As noted the axial trace of the dominant anticline is bent around the axis of a post-Belmont anticline with north-northwest–trending axial trace (Fig. 10).

The kinematic relations among the Bogles thrust, the dominant anticline, and the culmination are an important question. Did the anticline develop before, during, or after the emplacement of the Bogles allochthon at its present position? Did the culmination evolve concurrently with initial folding or later, perhaps during cross folding? We interpret the dominant anticline to be syn- or postemplacement and not to have been transported in the Bogles allochthon because of the anomalous deformation in the footwall (Anse La Roche Formation) of the Bogles thrust near Bogles (Fig. 10). If such deformation is really local, it is presumably related to the development of the suprajacent dominant anticline and (or) the anticline's culmination, implying one or both formed in place. On this basis, we interpret the dominant anticline to be a fault-propagation fold caused by out-of-sequence imbrication within the Bogles allochthon. Further kinematic constraints are the following. First, thrusting within the Bogles allochthon is recognized only in association with the Cherry Hill Basalt. The apparent restriction of such thrusting to the core of the dominant anticline may be attributed to the probably high flexural rigidity of the Cherry Hill relative to the suprajacent strata. At high stratigraphic levels of the dominant anticline east of the culmination, faulting and inclination of the axial plane are unrecognized. Second, the Bogles thrust has Eocene Belvedere Formation, not Cherry Hill Basalt, in its immediate hangingwall at Bogles, implying that the development of the dominant anticline and the culmination was complicated, certainly more so than simple imbricate stacking of initially horizontal layers. Third, even younger (Oligocene) Belvedere Formation exists in the hangingwall close to the Bogles thrust at High North (Fig. 10), a kilometer or so north of the culmination. This may be explained in one of two ways: (1) the hanging wall ramp of the Bogles allochthon as a whole intersects the thrust in northern Carriacou, and the tip of the Bogles thrust lies only slightly farther north (Fig. 16B-I); or (2) the development of the dominant anticline and (or) culmination caused stratal omission in an area north of the culmination (Fig. 16A-I).

Two models of fault-propagation origin of the dominant anticline are illustrated in Figure 16. Both employ the simplistic assumption that the allochthon initially contained horizontal plane parallel strata. The allochthon either continued north well beyond Carriacou (model A) or it had a tip in Carriacou (model B). Both models show a second stage in which an inclined fault propagation fold in the Belvedere Formation and an imbricate thrust below Cherry Hill Basalt develop in the fold's core (Fig. 16). The fold and thrust formed out of sequence within the allochthon, perhaps due to the onset of strong coupling across the Bogles thrust in the zone below the fold and north of the ramp in the Cherry Hill Basalt. The coupling caused Bogles thrust-parallel simple shear and deformation at the top of the footwall (Anse La Roche Formation). The third stage of both models then shows forward propagation of the imbricate thrust more or less along a flat above the coupled segment and farther north, back down to the Bogles thrust, either along a stratigraphic flat (model B) or a hangingwall ramp (model A). The third stage in both models satisfies the second and third constraints discussed above.

To conclude, the dominant anticline in the Bogles allochthon is perhaps a fault-propagation fold that formed out of sequence close to its current position. Its translation was northwesterly or southeasterly. The fold may have developed either during the late emplacement of the Bogles allochthon or after such emplacement, but before Belmont deposition. Model B (Fig. 16) implies generally northwesterly transport on the Bogles thrust because the model Bogles thrust tip existed in northern Carriacou. In contrast, model A provides no sense of thrust transport because the fault-propagation fold could have translated either forward or backward relative to allochthon movement.

The culmination of the dominant anticline may have developed progressively after initial folding portrayed in Figure 16. We interpret the Belmont A Formation to be folded at the eastern reach of the culmination (section Y-Y', Fig. 10), although substantially less than subjacent formations. Therefore, the out-of-sequence displacements may have continued into Belmont A time (probably early Miocene). Such late displacements are conceivably related to deformation that caused the superposed folding in the Belvedere vicinity (Fig. 10).

Younger contraction. All of the older rocks (Fig. 8b; Grand Bay Formation and below) of Carriacou contain structures that indicate horizontal east-west contraction of modest degree. The structures include thrust faults of small displacement and related cleavage in the Anse La Roche Formation, minor mesoscopic folds and a macroscopic antiform, all with north to northwest

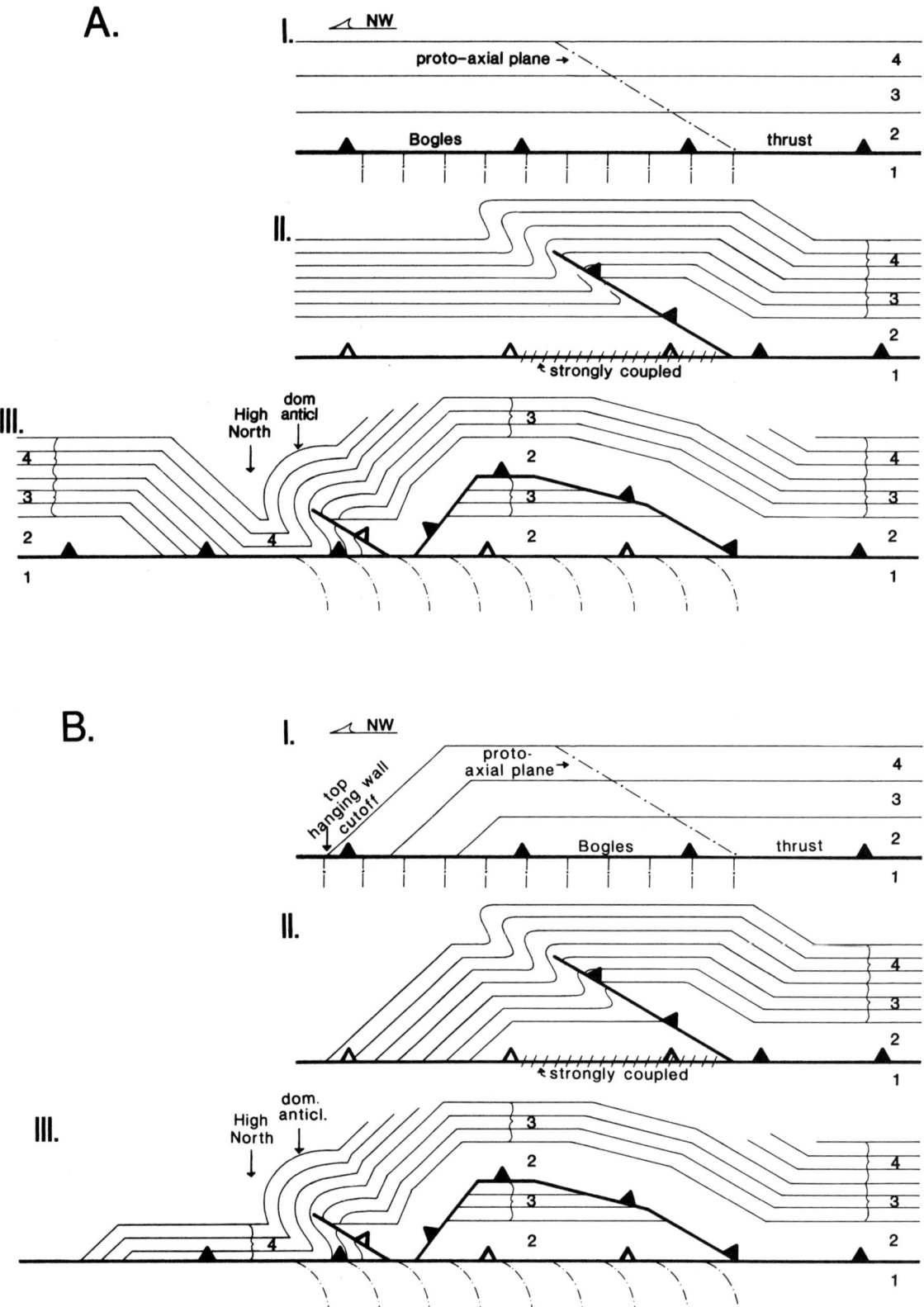

Figure 16. Alternative models, A and B, explaining development of the dominant anticline and thrust below Cherry Hill Basalt near Bogles village (Fig. 10), Carriacou, in timeseries I to III. Layer 1 is Anse La Roche Formation; 2 is Cherry Hill Basalt; 3 and 4 are Belvedere Formation; dash-dot lines are fold axial traces; lines with teeth are thrust faults—filled teeth when active, open teeth when inactive.

axial traces in the Belvedere Formation, and open folds with axial trends mostly between northwest and northeast in the Miocene sedimentary succession. In the Anse La Roche and Belvedere Formations, such structures are superposed on the early structures discussed in paragraphs above. Thus, it is reasonable to infer that the structures here named are contemporaneous and are younger than the middle Miocene formations (Fig. 8b), younger than about 15 Ma.

If it can be assumed that the generally north-south–trending Neogene dikes indicate horizontal east-west extension in the arc platform at about 11 Ma and thereafter, the Neogene contraction is constrained to a duration between about 11 and 15 Ma, middle Miocene. The younger contraction of Carriacou is probably local if arc magmatism in general indicates horizontal extension in the arc platform because early and middle Miocene arc volcanism occurred somewhere in the arc system to supply the Belmont and Grand Bay Formation sediments. On the other hand, the late contraction may have been regional and concurrent with magmatism in which case, its age is unknown after about 15 Ma.

As noted, it is possible that some of the deformation in the Bogles allochthon that caused the culmination near Bogles may have occurred in this late contractional phase of late folding about north-south–trending axes (Fig. 10).

Union Island

Union Island (Fig. 1) is underlain mainly by unaltered Neogene volcanigenic rocks (lava, debris flow, protrusion, dike) that are younger than 12 Ma, according to 8 K-Ar dates by Westercamp and others (1985) and Briden and others (1979). The eastern end of Union exposes older rocks (Fig. 17a) that we have studied chiefly in the area of Figure 17b where three units: sandstone-chert, massive porphyry, and breccia, are distinguished. The sandstone-chert unit is of particular importance because Westercamp and others (1985) said it is Cretaceous. We dispute their age and show below that such rocks are Eocene.

The sandstone-chert unit is entirely sedimentary and discussed at length below. It is Eocene, and has fault contacts with other units. The massive porphyry unit, called "ensemble volcanique altéré de Fort Hill" by Westercamp and others (1985), is a bleached, siliceous-appearing, massive feldspar porphyry that contains 30 to 40% fine to coarse altered feldspar phenocrysts and no preserved mafic grains. Its matrix is very fine grained at places where it has escaped alteration. The matrix contains local spaced foliation (Fig. 17b) but there is no deformation or preferred orientation of phenocrysts. This undated rock is interpreted to have been a viscous protrusion or shallow intrusion to explain its massivity and texture. Contacts between the sandstone-chert and massive porphyry units are brittle faults; both walls are brecciated, and bedding in the sandstone-chert unit is strongly discordant to the contact. The sandstone-chert unit does not lie above the massive porphyry as claimed by Westercamp and others (1985). At the roadside contact (Fig. 17b), however, the massive porphyry has relics of dark aphanitic matrix, suggestive of a chilled contact zone. This implies that the porphyry unit invaded the sandstone-chert unit and the fault is a faulted intrusive contact. The breccia unit, sparingly exposed (Fig. 17b), consists of brecciated foliated polymict breccia. Its clasts are vesicular lava, green aphanite, and altered rocks; the matrix is chalky feldspar and clay. Its relationship to the sandstone-chert unit is unknown.

The sandstone-chert unit (Fig. 17b) consists of fine and coarse volcanigenic sediments and of interbedded variably chertified limestone and mudstone. The three main lithotypes are:

1. *Sandstone.* This little-exposed lithotype contains thin and thick beds of fine- to coarse-grained, plane-laminated and massive-graded feldspar-volcanilithic sandstone. The sandstones are probably top-absent (Tab) turbidites or perhaps, grain flows.

2. *Thin-bedded cherty rocks.* These well-layered rocks occur as sets of thin beds up to at least 5 m thick in contact with either breccia or sandstone. The thin-bedded rocks are cherty to a degree that varies from vitreous to only slightly indurated and with abundant carbonate relics. The protoliths were probably fine-grained feldspathic sandstone, skeletal calcwacke, and marl that contains ghosts of radiolarians and of spotty, poorly preserved benthic and planktic foraminifers and nannofossils. The thin-bedded rocks probably represent pelagic and distal turbiditic deposits that have undergone varied and locally advanced silica diagenesis. The nearly complete destruction of radiolarians in these rocks implies they were a source of diagenetic silica.

3. *Sedimentary breccia.* The breccia contains angular fragments as coarse as 15 cm but mainly ¼ to 2 cm. Their compositions are feldspar porphyry, chert and other very fine grained rocks, and feldspar. Breccias are mostly clast-touching but grade to pebbly sandstone. They are generally structureless. Evidence for sedimentary origin is conformable interlayering with thin-bedded rock sections (Fig. 17b) and a mixture of igneous and sedimentary clast types. These are probably high-regime, channel-filling grain flows.

We found rare, poorly preserved nannofossils (*Sphenolithus moriformis* and *Coccolithus floridanus*) in one (specimen 10i, Fig. 17b) of nine specimens of the sandstone-chert unit examined from thin-bedded rocks with evident carbonate and *Globigerinatheka sp.* in several thin sections. These all have middle Eocene first appearance. The *Globigerinatheka* restricts the permissible age range from P10 to mid-P16. The only large foraminifers in our specimens are long-ranging *Amphistegina*. Radiolarian relics exist in several specimens but are impossible to identify owing to dissolution and recrystallization.

Westercamp and others (1985) obtained two specimens of the sandstone-chert unit from roadside exposures (Fig. 17b) that contained badly preserved bivalves and larger foraminifers. The specimens were said also to include typical Campanian nannofossils (Westercamp and others, 1985). The forams are *Amphistegina,* as in our specimen (J. Butterlin, oral communication, 1986). The nannofossil species named by Westercamp and others (1985), however, are completely different from those in our specimen 10i. The conflict has been resolved with the cooperation of those authors. The original slides of particulate material from their two specimens do indeed contain well-preserved Late Cre-

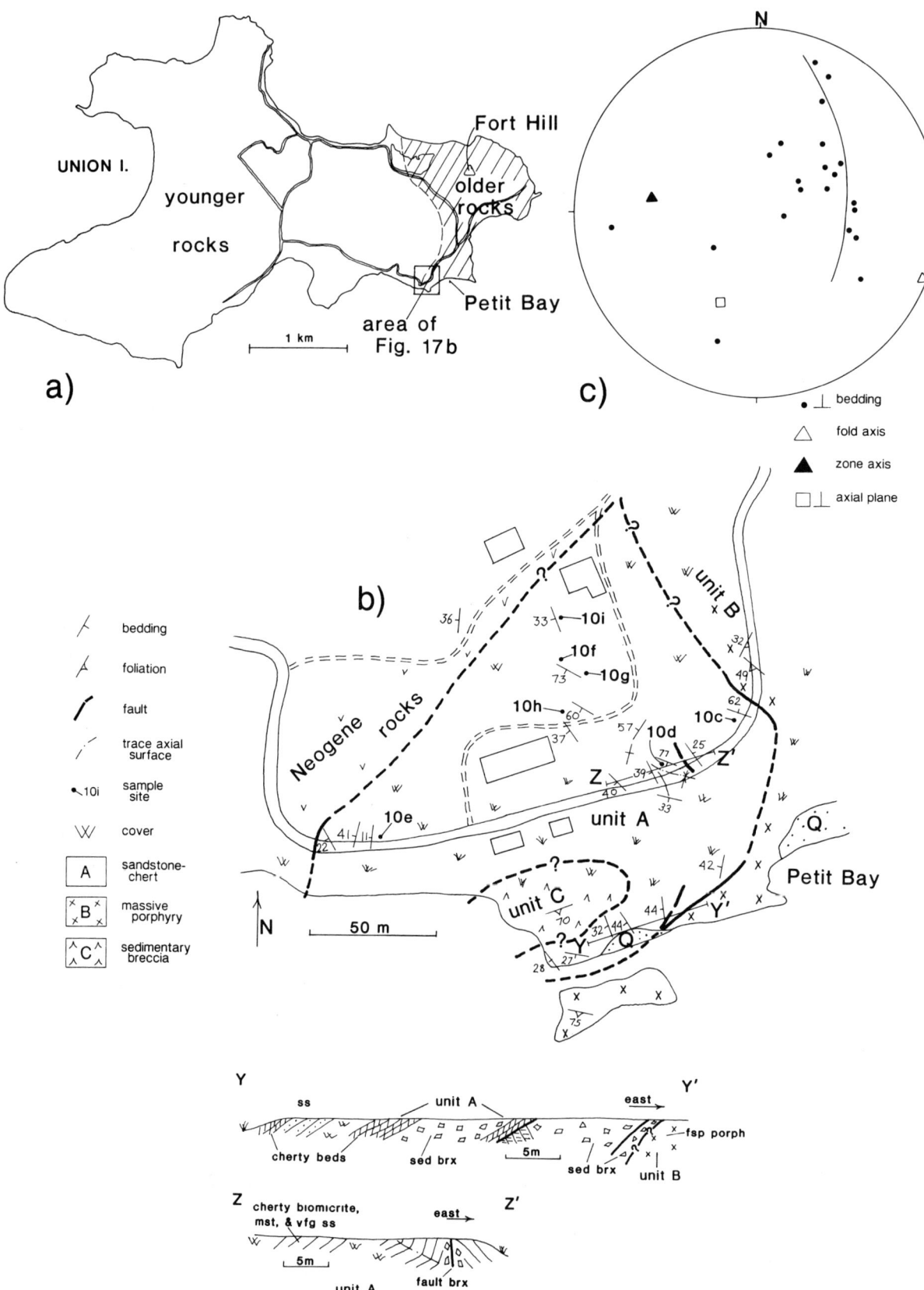

Figure 17. a) and b) Geologic maps and sections of Union Island; c) orientation data for sandstone-chert unit.

taceous nannofossils, but new slides from the same specimens prepared by them and one of us (KPN) contain no nannofossils at all. This is consonant with the advanced diagenesis of their samples. Therefore, the Cretaceous nannofossils must be laboratory contaminants introduced during preparation of their first (1984) slides. As a result, Mesozoic rocks have not been discovered on Union Island or elsewhere in the southern Lesser Antilles.

The association in the sandstone-chert unit of high flow regime volcanigenic sediments together with interbeds of probable pelagic origin suggests deposition in a basinal setting with inflow of clastic sediment by sediment-gravity flows. The dated specimens are pelagic lithotypes and thus, probably record true age of deposition.

Bedding in the sandstone-chert unit varies substantially in orientation (Fig. 17c), implying folding. Only one fold hinge was found, however. This is probably due to the lack of outcrop, but also to faulting of sections with different layer attitudes (sections, Fig. 17b), implying that folds are broken. The continuous fold is close and has a north-dipping axial plane and subhorizontal axis (Fig. 17c). Bedding as a whole in the sandstone-chert unit is in a diffuse girdle with a west-plunging axis, together with several highly aberrant attitudes (Fig. 17c). Domainal analysis indicates local zonal axes between northwest and southwest trends. The strongly aberrant attitudes can be explained by local rotations due either to local landsliding or faulting. In general, the bedding and fold orientations indicate north-south contraction. The single fold suggests southerly vergence. There is no cleavage or evidence of dissolution or metamorphism during deformation.

Prune Island

Prune Island (Fig. 1), also known as Palm Island, exposes undated deformed sedimentary rocks in locally good outcrops through Quaternary cover. Such outcrops afford the best structural control of the central Grenadines (Fig. 18). No magmatic rocks crop out. The sedimentary rocks are chiefly sediment-gravity flows, including turbidites, and comprise two compositional types: volcanigenic and mainly nonvolcanigenic (Fig. 18). Both types contain minor interbedded pelagic rocks. At the two exposed depositional contacts between strata of the two types, the nonvolcanigenic beds are conformable and below (Fig. 18).

The volcanigenic successions (Fig. 18) are coarse grained, commonly massive but locally well bedded and graded. Boulder conglomerate and sedimentary breccia, generally unsorted and massive, occur both in tabular layers and channels. The clasts are all angular and composed of dark aphyric and porphyritic basaltic rocks. These are interlayered with volcanilithic-feldspar-pyroxene sandstone, locally pebbly, that is mainly in sets of top absent, upward-fining thin and thick beds. Bouma zonations are Ta, Tab, and Tac. At the southeastern point (Fig. 18), such sandstone grades up to calcarenitic tops (Tb), and discrete beds of chert, 3 cm thick, occur between some sandstones. The volcanigenic beds are notably free of mud and mudclasts in contrast to the nonvolcanigenic ones.

The nonvolcanigenic successions include the following layer types: (1) granule conglomerate and sedimentary breccia that is thick bedded and massive; sedimentary breccias have clasts as coarse as 20 cm; breccias are both tabular and scour-filling; (2) well-graded sandy turbidites; thin to thick bedded and fine to very coarse grained; they include Tabc, Tabce, Tbc, Tbce, Tce zones and are commonly pebbly with outsized mudclasts, some of which show pliability during sediment transport; (3) ungraded, thin-bedded sandstone and mudstone; and (4) thin layers of variably chertified, fine-grained rocks with microfossiliferous pelagic protoliths that were probably marl, mudstone, and sandstone; these occur as gradational tops to beds of type 2 and as discrete interbeds with sandstones of type 3. The coarse sand and coarser clasts are fragments of fine-grained sandstone, partly chertified rocks, and chert that is gray, green, or red. The composition of finer particles is unknown. The nonvolcanigenic beds include classic turbidites and related proximal deposits and products of interturbidite pelagic downfall. They also include the nonturbidite types of type 3, above; such sandstones, however, may represent bottom-current reworking of former thin-bedded turbidites. Diagenesis of the pelagic layers and adjacent fine-grained sandstone is varied but generally well advanced such that sections include both moderately soft muddy rocks as well as tough chert. The mudclasts that probably came from channel-wall collapse show a similar spectrum of diagenesis. This suggests that at least much of the fragmental content of the nonvolcanigenic beds is reworked from similar, penecontemporaneous deposits.

We attempted without success to recover microfossils from softer pelagic layers and mudclasts in the nonvolcanigenic beds.

The Prune Island beds evidently accumulated in a basin that was fed by turbid flows from disparate source regions, one volcanic, the other sedimentary. The volcanigenic beds represent mainly upper-fan or upper midfan facies whereas the nonvolcanigenic ones contain upper- to lower-fan facies. It is not clear whether the two delivery systems operated simultaneously or sequentially. Because the nonvolcanigenic beds contain much recycled material that is similar to their own primary deposits, their source may have been uplifted earlier sediments of the same succession in a nearby area. This suggests tectonism during sedimentation.

The structure of the strata of Prune Island is deduced mainly from outcrops at the western and eastern sides of the island (Fig. 18, sections a, b). The mainly nonvolcanigenic strata of the westernmost area are in partly faulted upright close major folds with probable southerly vergence. The vergence is interpreted from different lengths and dips of the limbs of the major anticline. Figure 18a shows that three fold hinges (2 major, 1 minor) plunge shallowly east-northeast, about parallel to the zonal axis for the western area. Bedding on the north limb of the anticline, however, deflects from the zonal girdle as traced west to east, parallel to the zonal axis (Fig. 18a). This smooth deflection is interpreted to be due to an open second syncline with axis that locally plunges northeast. The second fold was not detected on the south limb of the first anticline, probably because of lack of exposure.

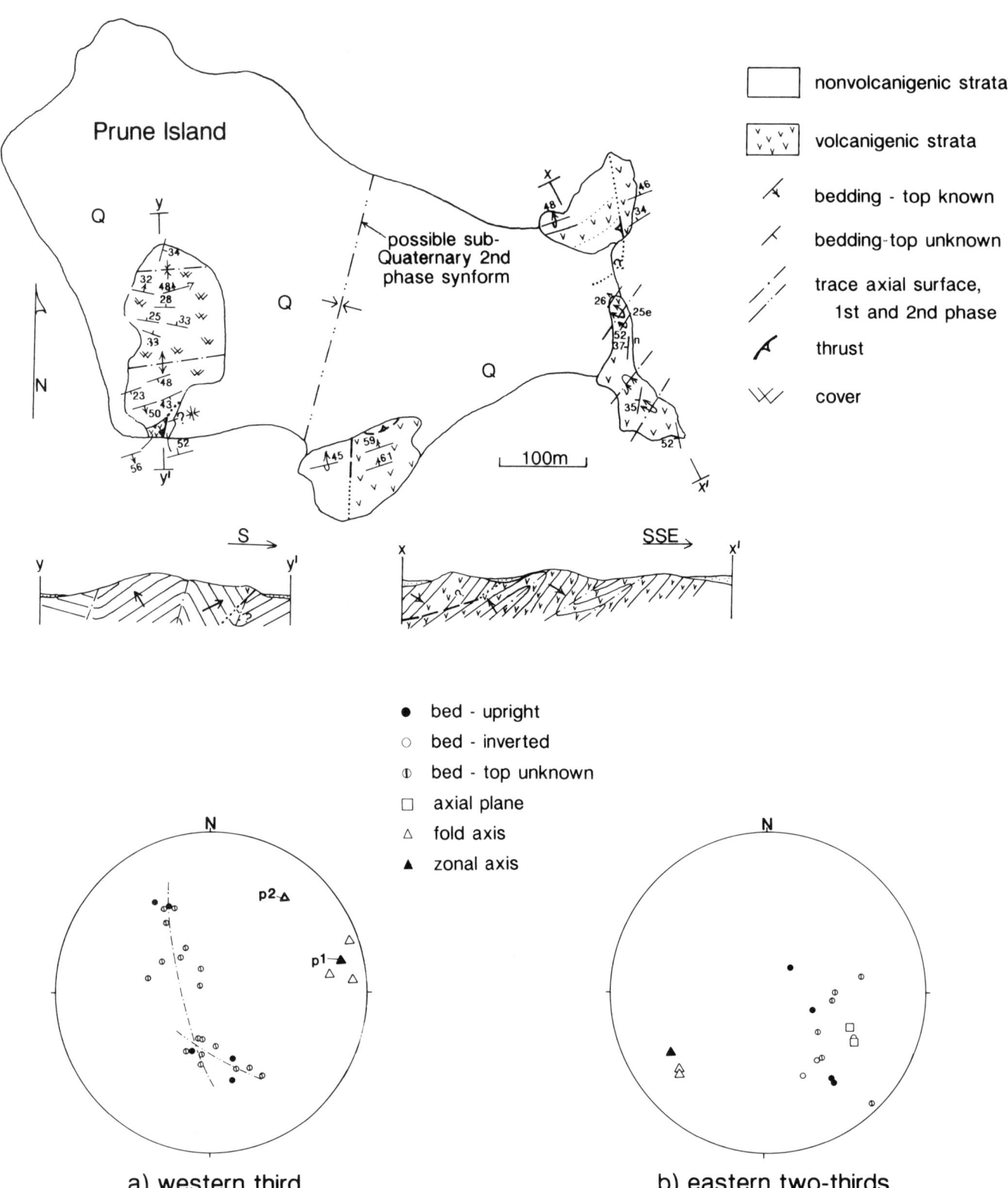

Figure 18. Geologic map, sections, and structural orientation data of Prune Island.

The eastern outcrops show greater deformation than the western, by virtue of tighter major folds and juxtaposition of two sections on a west-dipping, low-angle fault. Two major anticlines are exposed, each overturned to the south and each with an interlimb angle of about 50°; an intervening syncline is inferred (Fig. 18b) in an area of poor exposure. Orientations also differ in the eastern and western areas; in the eastern, axial traces trend N30°E, and axes of both individual folds and the bedding girdle plunge shallowly west-southwest (Fig. 18b).

The contrast in structure between the two areas can be explained either by rotation due to superposed deformation or by heterogeneity developed during progressive deformation probably accompanying thrusting. The first is likely because (a) there is evidence for open second folding in the western area, (b) the first-phase fabrics are reasonably systematic in each area and not indicative of large heterogeneity, and (c) the rocks in each area seem to belong to the same succession. Further, the two fabrics (Fig. 18a, b) can be brought into coincidence by rotation about a northerly trending axis. This suggests that an open major second phase synform with approximately north-striking axial plane lies in central Prune Island (map, Fig. 18). The second explanation is possible because thrusts of varied orientation cut the layered rocks, and orientations of strain in hanging and footwalls may reflect local conditions of ramp attitude and ramp failure.

In general, the undated strata of Prune Island record early shortening with a horizontal trend between ESE-WNW and SSE-NNW, together with southeasterly overriding. The early structures either were later rotated during second folding with northerly axial trace or developed somewhat heterogeneously to begin with.

Jamesby

This 200-m-long island (Fig. 1) exposes a continuous succession of volcanigenic sediment gravity flows and turbidites with calcareous pelagic interbeds that occupy a major syncline overturned to the southeast (Fig. 19). Westercamp and others (1985) dated a bed on Jamesby as late middle Eocene.

The predominant lithotypes are massive pebble and granule sedimentary breccia and pebbly sandstone, all with much matrix and poor clast orientation, in beds 1 to 10 m thick. Clasts are mainly feldspar microporphyry, but quartz-bearing volcanic clasts, rare in older rocks of the Grenadines, also occur, as noted by Westercamp and others (1985). Variably chertified mudstone and marl also occur as clasts. The coarse rocks are interbedded with sandstone and calcareous mudstone that exist in successions 0.01 to 10 m thick. Sandstone is quartz-bearing and volcanigenic, and thin to thick bedded. Mudstone occurs mainly as thin-graded tops of sandstone beds. It is moderately chertified in the overturned limb (Fig. 19) but not in the upright limb. Turbidites are mainly thin and medium beds with sandy or pebbly bottoms and muddy tops whereas thick breccias and sandstones are little graded and are probably debris flows and grain flows, respectively.

Planktic foraminifers (Table 2) and nannofossils recovered by Westercamp and others (1985) from a limey interbed indicate a composite age of P12 (*M. lehneri*) or NP16, late middle Eocene, between about 43 and 45.5 m.y. P. Andrieff (written communication, 1990) informed us that the *Globorotalia pomerli-cerroazulensis* (s.s.) group is well represented in the Jamesby bed, in contrast to specimens from Baradel, which contain only ancestral taxa of the *cerroazulensis* lineage. This indicates the Jamesby bed is clearly younger than the Baradel chert, by as much as about 6 m.y. (P11 versus P12).

Tectonic structures in the Jamesby rocks are of three types: major syncline, thrusts and related folds, and foliation. The major syncline has a half-wavelength, >200 m, encompassing all exposed beds (Fig. 19). It is overturned to the south-southeast and has an axis plunging about 20° northeast (Fig. 19a). The overturned limb is cut by many faults, mainly thrusts where the sense of slip is known. These cut subhorizontally across bedding and commonly splay into bedding subparallel faults. The nonplanar faults cut parts of the overturned limb into series of phacoidal blocks, much as in the chert unit of Baradel. Such blocks contain open folds of varied orientation (Fig. 19b) that are accommodations of bedding to changes of fault attitude.

Spaced foliation of generally steep dip (Fig. 19a) occurs in the matrix of some sedimentary breccias. These may be axial planar to the major syncline, but few foliations measured dip more steeply than the major fold's bisecting plane (Fig. 19a). Layer subparallel foliation exists in some unchertified muddy beds and is folded in the fault-accommodation folds. Such foliations may represent strains taken up early in bedding-parallel faulting.

The main occurrence of thrusts in the overturned limb of the syncline suggests they were generated late in the folding or after folding and represent failure of the misoriented steep limb under progressive subhorizontal simple shear.

Baradel

Baradel (Fig. 1) is underlain by deformed sedimentary rocks of two dissimilar lithic units (Fig. 20) that are joined by a thrust, not a depositional contact as thought by Westercamp and others (1985). The structurally higher chert unit consists of cherty pelagic and distal turbiditic beds and is dated early middle Eocene by Westercamp and others (1985). The lower unit contains coarse arc-derived volcanigenic sediment and is undated.

Rocks. Chert unit. The chert unit contains characteristically pale green thin beds that have undergone chertification to varied degrees. Protoliths were apparently calcareous biomicrite, micrite breccia, radiolarian micrite or marl, and turbidite. Turbidites are mainly very fine grained sandstone that fines upward to mudstone, both now much replaced by silica. They are base-absent except for a single thin bed of plane-laminated sandstone that includes floating volcanic granules. The composition of the sand and granules is unknown. No bottom or top of the unit is exposed. The stratigraphic thickness of rocks in outcrop is uncertain owing to deformation, but between 10 and >40 m. The unit is locally foliated.

As noted earlier, the chert unit of Baradel may be of similar composition to and correlative with allochthonous cherts of Mayreau.

Volcanigenic sediment unit. The unit of volcanigenic sediments contains mainly massive, unsorted granule- or pebble-rich sedimentary breccia, together with lesser stratified conglomerate and sandstone. Fragments are entirely volcanic and of arc origin: fresh or altered feldspar porphyry, fine-grained holocrystalline lava, and minor foliated rock. A few fragments are rounded whereas most are angular. The stratified rocks are medium-grained sand or coarser and commonly plane laminated. The unit is not foliated. The sedimentary breccia is widely cut by mafic dikes and nonsedimentary breccia with clasts of local breccia. The secondary breccia is probably diatremal.

Dating. Our studies of thin sections of cherty micrite together with investigations by Westercamp and others (1985) of the Baradel chert unit found ten species of planktic foraminifers (Table 2). We recovered no radiolarians or nannofossils from this unit, apparently because such material dissolved during diagenesis. P. Andrieff (written communication, 1990) revised identifications by Westercamp and others (1985) to exclude *Globorotalia pomeroli* (Table 2). The remaining nine foram species give a zonal age in the *Globigerinatheka subconglobata* zone (P11), equivalent to the nannofossil NP15 zone, early middle Eocene and about 45.5 to 50 Ma. No fossils have been found in the volcanic sediment unit.

Deposition. The chert unit is certainly an open marine basinal deposit as indicated by the thin limestones that contain planktic faunas and by their intercalation with thin muddy turbidites. Moreover, the extensive chertification implies the former

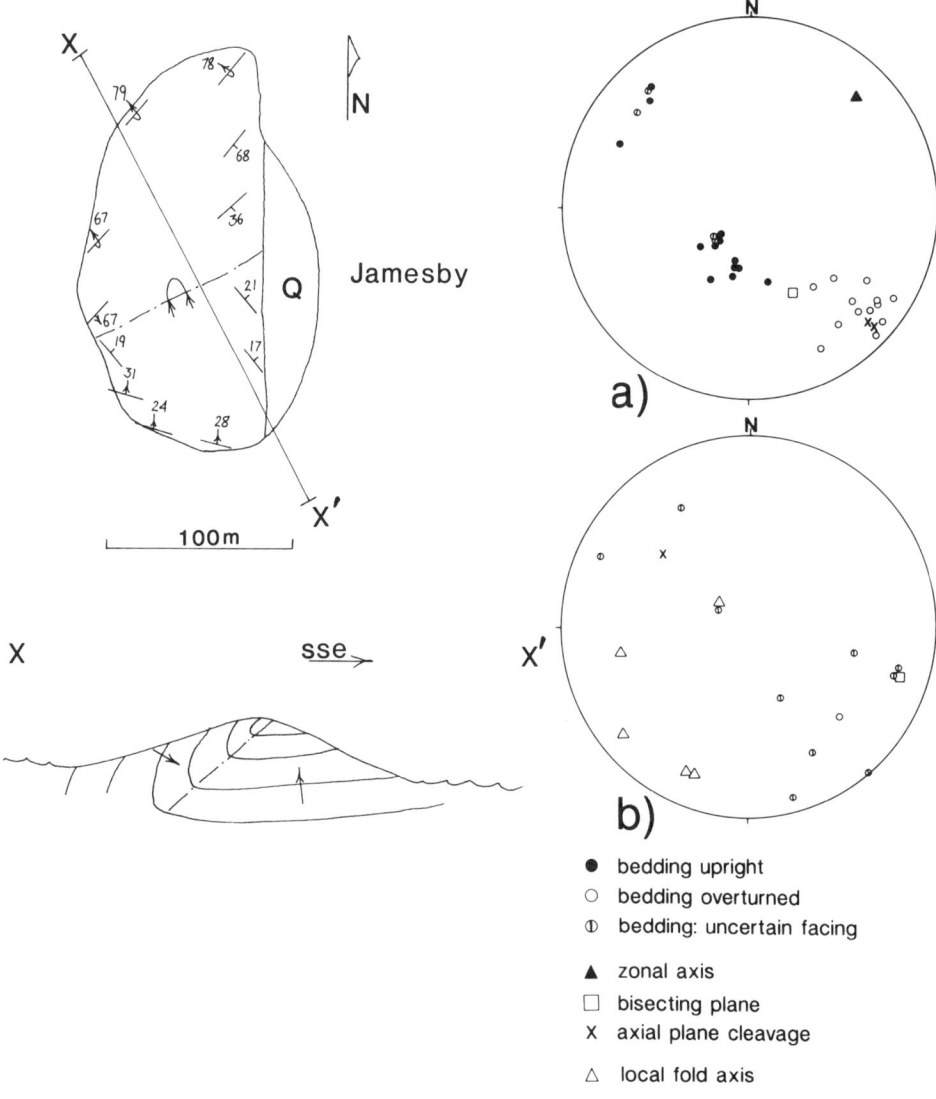

Figure 19. Geologic map, section, and structural orientation data for Jamesby. Diagram a) shows attitudes related to macroscopic syncline. Diagram b) shows attitudes in phacoidal blocks in disrupted overturned limb of macroscopic fold.

existence of much biogenic silica that was probably initially in planktic organisms. The siliciclastic grains of the chert unit are probably all volcanigenic, but their composition is unknown, due to diagenesis. Therefore, it is uncertain whether such sediments were from a magmatic arc or from basalt of a nonarc setting. The volcanic sediment unit is arc-derived and probably represents a submarine channel fill by virtue of its massivity and coarseness.

Structure. Although bedding in the chert unit includes a homoclinal component with northwest dip, the unit is in fact disrupted by a network of faults that are mainly layer-parallel but locally cut across layers (section X–X', Fig. 20). The faults juxtapose lenticular packets of beds against one another. Displacements are unknown.

Many layers of the chert unit contain a first foliation that is parallel or at low angles to the bedding. The first foliation, best developed in carbonate protoliths, is spaced and wavy. It is folded together with bedding (Fig. 20). We found no early folds to which this foliation is axial planar, although one tight minor fold with northwest-plunging axis (Fig. 20) in unfoliated beds could be an early fold.

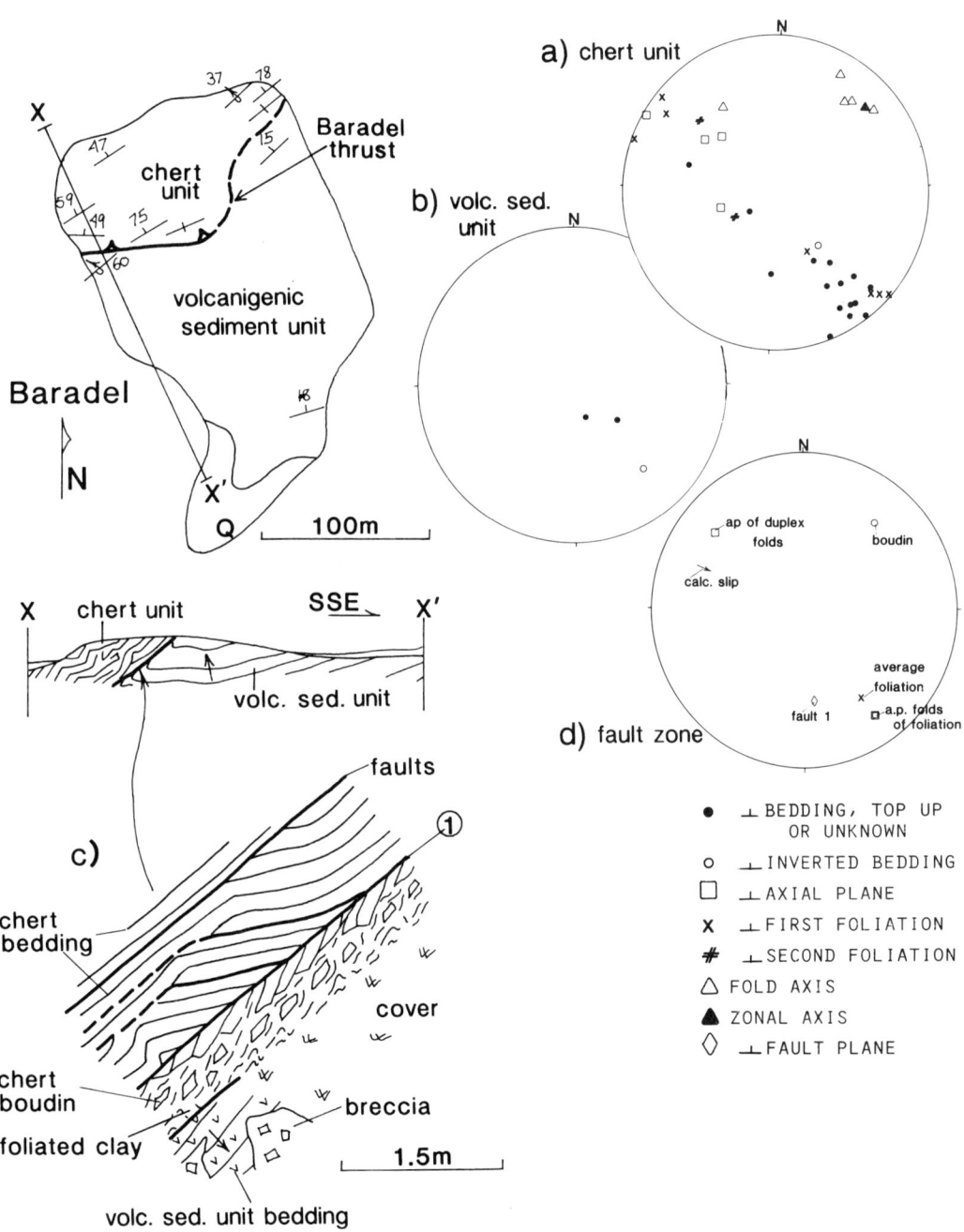

Figure 20. Geologic map and section for Baradel; a), b), and d) give structural orientation data; c) is a local section in the Baradel thrust zone.

The most evident structures are open to tight folds of bedding and first foliation (section, Fig. 20), with mainly chevron but locally concentric hinges. These have northeast-plunging axes and southeast-dipping axial planes and local axial plane (second) foliation (Fig. 20). Such folds are probably synfaulting and caused by accommodation of layering attitudes to ramps (open folds) and by heterogenous strain in fault zones.

The structure of the volcanigenic sediment unit is poorly known due to its paucity of layering. The few beds suggest that it is a northwest-dipping homocline except just below the thrust with the chert unit where the beds are dragged and inverted (Fig. 20, section X–X′, and b). The partial girdle to beds is about homoaxial with the fold and zonal axes in the chert unit. This suggests that faulting within the chert unit was concurrent with its thrust transport above the volcanic sediment unit.

Features of the thrust zone are exposed above the beach on the west side of Baradel (Figs. 20c, d). Fault 1, at the base of the chert unit, is planar and dips north-northwest. Between it and a higher parallel fault is a duplex of chert beds. Below it are fault rocks in a 0.5-m-thick zone above another subparallel but poorly exposed fault and below that are inverted beds of the volcanigenic sediment unit. The fault rock consists of imbricates and boudins of chert from the hangingwall in a matrix of foliated clay (Fig. 20c). At the base of the fault rock, the foliations are in southeast-verging microfolds (Fig. 20c). The boudin axes parallel fold axes higher in the chert unit, and the axial planes of folds in the duplex have similar orientations to those higher up. Assuming that in the fault rocks the intermediate principal strain paralleled boudin axes and that maximum shortening was normal to average foliation, the calculated thrust slip trends S70°E; the sense is corroborated by vergence of microfolds of foliation.

The bedded rocks in the footwall are invaded by discordant bodies of breccia of volcanic breccia. We suggest these were mobilized by increases of fluid pressure in the volcanic sediment unit as it was overridden by the chert unit.

Sequence of events. The evolution of the chert unit of Baradel is interpreted as follows: (1) deposition in a basinal site with predominant pelagic downfall and moderate influx of fine volcanigenic grains from an unknown source in early middle Eocene time; (2) development of first foliation in at least some beds probably before or during extensive silica diagenesis; (3) diagenesis or low-grade metamorphism, producing layer hardening and early dissolution foliation; and (4) internal faulting that led to folding and disruption, possibly related to thrusting east-southeast over the volcanic sediment unit. Events 2 and 3 were probably early whereas the timing of 4 is unknown.

Mayreau

Mayreau (Fig. 1) exposes three units of older rocks: the here-named Mayreau Basalt and Anse Bandeau Formation, and an allochthonous unnamed chert unit (Fig. 21). The named formations are the oldest dated rocks in the southern Lesser Antilles arc platform. Mayreau also contains widespread intrusive igneous rocks of mainly uncertain but probably Neogene age. The number of intrusive events among the Neogene rocks probably exceeds four.

Mayreau Basalt. The oldest rocks on Mayreau are pillow basalts that occupy a moderately deformed, generally south-dipping and south-facing succession (Fig. 21, unit Ep). At least 500 m of Mayreau Basalt crops out below the Anse Bandeau Formation, assuming that basalt is not repeated by covered or unrecognized faults. The base is unexposed. The Mayreau Basalt is not directly dated but is assumed to be Eocene because the concordantly overlying (but faulted) Anse Bandeau Formation includes similar pillow lava and pelagic limestone of Eocene age (Fig. 22a). We dated two specimens of Mayreau Basalt by K-Ar (Fig. 21, Table 1), but the values, 12 and 14 Ma, probably record the age of Neogene intrusion and related alteration, certainly not the age of extrusion of pillow basalt.

Mayreau Basalt is mainly a pile of ropy pillows of widely varying and commonly large aspect ratio (length/thickness 1 to 7, commonly 4 to 5). Local concentrations of calcareous red sediment occur in pillow successions. No discrete lava flow units can be identified within the sequence. The pillow lavas are plagioclase-clinopyroxene basalts that are strongly amygdaloidal and mainly aphanitic but locally finely porphyritic. They are moderately to strongly altered by calcitization and hydration and contain few to moderate numbers of carbonate and carbonate-quartz veins. The lavas are not foliated.

In thin section, the pillow lavas vary from nonporphyritic to finely porphyritic to glomeroporphyritic. Each has similar groundmass petrography: felted chlorite + ironoxide + carbonate + feldspar. These are almost certainly products of devitrification and hydration of basaltic glass. Much of the carbonate may be postdevitrification as indicated by its large variation in concentration (up to 80% of the groundmass) and the widespread occurrence of carbonate veins. Groundmass feldspar is needley and commonly swallowtailed. The principal phenocryst type is plagioclase; the size and sphericity of such phenocrysts increase with phenocryst abundance. Where there is no agglomeration, phenocrysts are solely plagioclase. Phenocryst clusters, however, may or may not include clinopyroxene. In glomeroporphyritic rocks, clinopyroxene is commonly ophitic with respect to lath-shaped plagioclase phenocrysts. Amygdules are composed either of chlorite or carbonate.

On the bases of major and minor element and rare-earth compositions of whole rocks and of clinopyroxene phenocryst compositions, the pillow basalts are nonalkaline and interpreted to be of spreading rather than island-arc origin (Speed and Walker, 1991). They are more likely of backarc than midoceanic origin, mainly from regional structural considerations (Speed and Walker, 1991).

Westercamp and others (1985, their Fig. 13) claimed that tuffs are widely interbedded with the pillow basalt. However, the rocks they called tuff in the well-exposed sea cliffs north of L'Anse Bandeau (Fig. 21) are in fact altered dikes, sills, intrusive breccia, and diatremes. These clearly postdate the basalt by virtue either of included clasts of basalt, feldspar porphyry, and microdiorite and of the existence of relict chill margins and boundaries that crosscut the pillow lavas, and are all probably Neogene. The K-Ar dates of Mayreau Basalt (Fig. 21, Table 1) indicate resetting in late Miocene time and probably, the age of the invading rocks.

Anse Bandeau Formation. This is a 35-m layered sequence that overlies concordantly the Mayreau Basalt with fault contact (Figs. 21, 22). The Anse Bandeau (units Es and El, Fig. 21) consists of red pelagic limestone, basalt, and basaltic sediments. The clastic sediment occurs in the debris flows, turbidites, and massive sandstones (Table 4). The sand is plagioclase-lithic, and gravel consists of granules to boulders of mainly vesicle- or amygdule-rich basalt. The beds are 0.25 to 5 m thick.

The red limestone near the base (unit Esi, Figs. 21, 22, Table 4) was dated with forams by Westercamp and others (1985, their Fig. 13, sample GR72), giving the same zonal age range (see below) as the composite age of the limestone at the top of the Anse Bandeau Formation (Fig. 22a). The formation includes above the dated lower limestone a thick pillow lava (Esd) and a basaltic sill (Ese) that is intruded at the pillow lava's base. These are the highest and youngest Eocene magmatic products, respectively, on Mayreau. The pillow lava is identical in pillow morphology and pillow and matrix petrography to those of the subjacent Mayreau Basalt. From this, it is assumed that Mayreau Basalt is also Eocene and that the pillow lava of the Anse Bandeau Formation was the final extrusion of nearly continuous basaltic effusion.

Well-indurated red biomicrite composes the highest exposed beds of the Anse Bandeau Formation (unit El, Figs. 21, 22; Table 4). The unit includes as much as 4 m of thin limestone beds below a top that is variably a fault, intrusive contact, and unconformity with Neogene cover. The rock consists of finely crystalline carbonate matrix and partly replaced planktic forams that have no stratigraphic concentration. The degree of diagenesis varies among layers as indicated by variable hardness, and this may be influenced by widespread hydrothermal alteration that proceeded out from cracks at high angles to bedding. The limestones are locally laminated and bioturbated. Bioturbation marks are chiefly horizontal burrows, but some cut bedding. The limestone is almost certainly a pelagic deposit as indicated by its lack of particle sorting, faunal composition, and continuity with basinal strata of subjacent units within the Anse Bandeau Formation.

A middle or lower Eocene age for the upper red biomicrite was first proposed by J. B. Saunders and H. Bolli (Martin-Kaye, 1969) from the identification of *Globorotalias* in thin section. Later, Westercamp and others (1985) identified foraminifers in thin sections that give a composite age of the *Hantkenina nuttali* (P10) and the *Globigerinatheka subconglobata* (P11) zones (Table 2). Patrick Andrieff (written communication, 1990) and we have examined old and new thin sections of the upper red micrite (unit El) of the Anse Bandeau Formation, confirming earlier identifications. We find a particularly rich fauna of *Globigerinatheka sp.*, which implies the assemblage is no older than middle *H. nuttali* zone. Our search found only a single nannoflora species, *Discoaster barbadensis,* and no identifiable radiolarians. Such ages indicate deposition in early middle Eocene time, between about 46 and 50 Ma. The similar fauna in limestones at top and bottom of the formation imply the formation's age is likely within this range.

The limestone contains spaced cleavage that is poorly preserved, probably because of hydrothermal alteration. It also contains shear bands of 1 to 5 cm thickness and joints.

Chert. Thin-bedded cherty rocks occur in a small outcrop area at Anse Bandeau (C, Fig. 21b). They lie above unit El of the Anse Bandeau Formation with low-angle fault contact (Fig. 22, Table 4). The cherty rocks and the fault zone are intruded by microdiorite of probably Neogene age. The cherty rocks are undated, but they are probably pre-mid-Miocene because they were deformed and faulted in place before igneous intrusion and because of their compositions, as discussed below.

Layer compositions of the chert (Table 4) imply that protoliths were thin beds of varied composition, including siliceous biogenic beds, carbonate of possible depositional origin, and fine-grained siliciclastic beds. Their position and probable partial pelagic origin suggest such beds are correlatives of the Anse Bandeau Formation, perhaps originally from strata higher than biomicrite of unit El. The cherty beds of Mayreau are like those of the middle Eocene chert unit of Baradel about 4 km east. The similarities are layer character and the varied (but poorly defined) protolith particle compositions. It is possible, therefore, that the two chert units are correlative and that the Baradel chert unit and the sedimentary succession at Anse Bandeau were continuous.

The cherty beds at Anse Bandeau are broken into phacoidal slices, each 0.1 to 1 m thick and as long as >3 m. Beds within the phacoids are openly folded. No foliation is evident.

Interpretation of older succession. The lithologic sequence of older rocks of Mayreau suggests the following evolution. The main succession of pillow basalt formed by spreading in an open-marine environment that was isolated, either by distance or barriers, from sediment of continental or arc provenance. The overlying limestone bed (Esi) indicates a lull in extrusion in early middle Eocene time. The succeeding clastic rocks (unit Es, Anse Bandeau Formation) include deposits of coarse basaltic debris that probably point to the development of a local edifice of pillow basalt, perhaps a fault scarp, late in the basaltic magmatic event. The last extrusion occurred during the subwavebase accumulation of the basaltic clastic sediments. In the succeeding duration, the basaltic succession was covered by pelagic sediment (unit El) in an environment still isolated from other sediment types. The pelagic limestones imply that at least the latest pillow basalts developed at or above the carbonate compensation depth, which was 3.5 km in mid-Eocene time (Kennett, 1978). The bedding-parallel burrows and absence of benthic megafossils in the limestone suggest, however, bathyal or greater depths. If the allochthonous cherty rocks of Mayreau are correlative with the Baradel chert unit and were originally continuous with the Anse Bandeau Formation, it could be inferred that the basin floor subsided after deposition of the biomicrite of unit El.

a.
Saltwhistle Bay
X
X'
N
Q
u
Ep
Lady Rock
Tarzan
MAYREAU
Station Hill
KAr=6.4
Saline Bay
Windward Bay
Upper Bay
Lande-Ci
area of fig. b)
Es
El
c
12°38.5'N
12°38.0'
61°24.0'W
61°23.5'
0.5 km
Q Quaternary sediment
u Quat. sediment and pre-Quat. rocks undifferentiated
younger rocks, Neogene and Neogene?
N undifferentiated igneous rocks
volcanic breccia
microdiorite
Anse Bandeau Fm.
El upper limestone; Eocene
Es turbidite, lava, limestone; Eocene
Mayreau Basalt
Ep pillow lava; Eocene?
c chert
low angle fault
bedding, no tops
bedding, top known
cleavage
dike
SSE
SL
m
100
200
250m

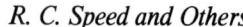

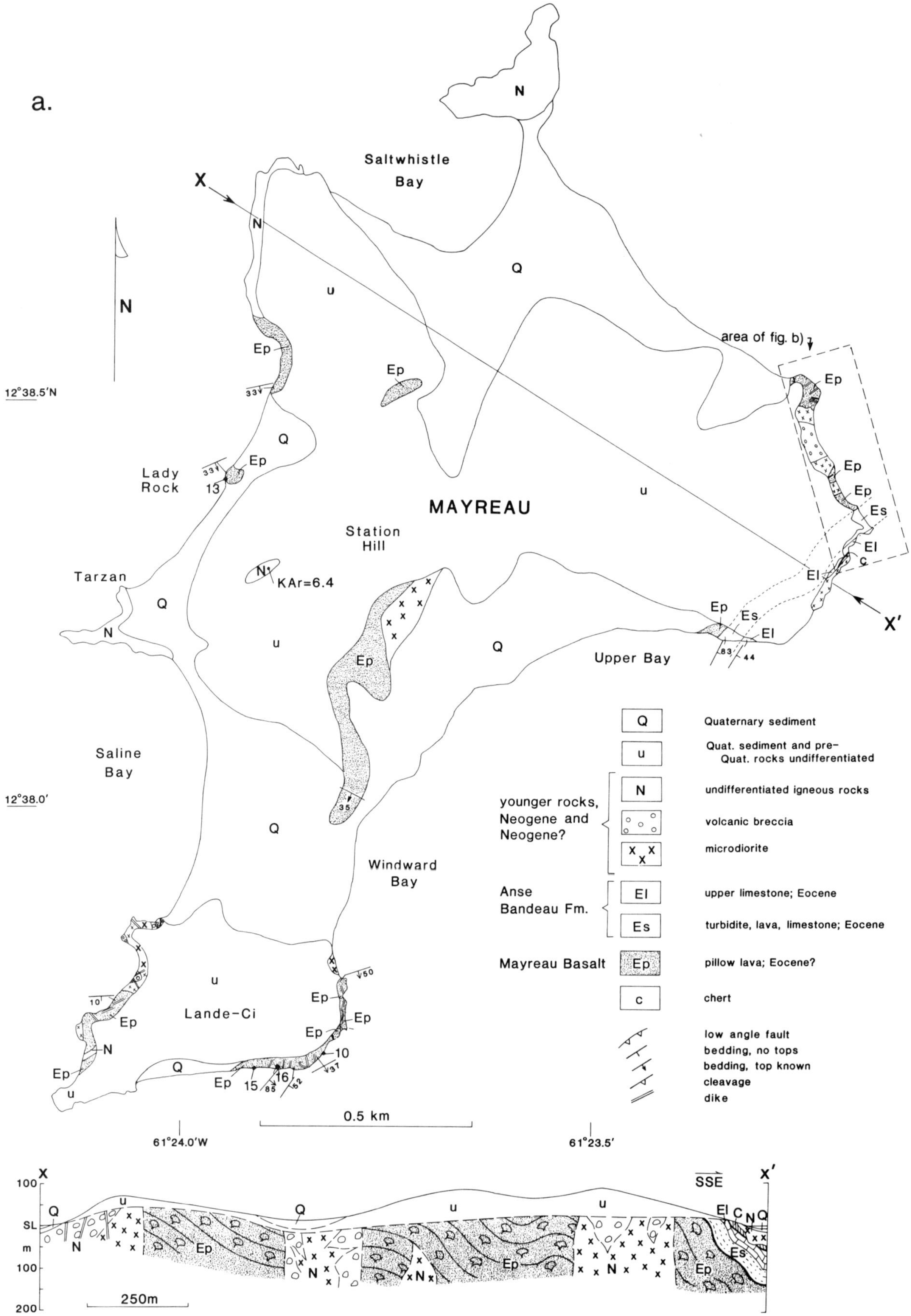

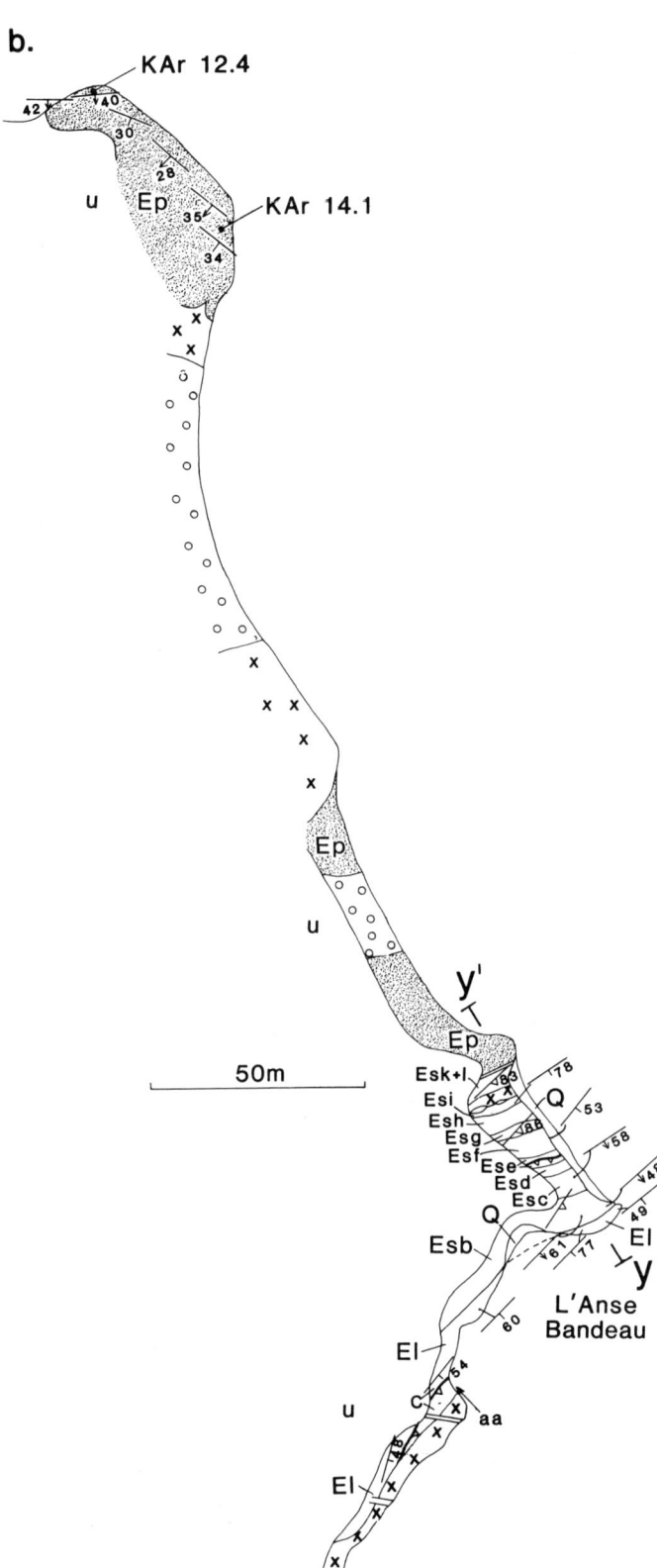

Figure 21. a (on facing page), Geologic map and section x–x′ of Mayreau; b, enlargement of coastal strip at and northwest of L'Anse Bandeau. Section y–y′ on Figure 22.

Younger rocks. Lithic units of Mayreau that are younger than units discussed above are microdiorite, breccia, and undifferentiated rocks (unit N, Fig. 21), which are mainly dikes in high concentration. The units are differentiated at coastal exposures (Fig. 21), but they are lumped together with Quaternary(?) mudflows and other sediments at most inland sites (unit u, Fig. 21) because exposures do not permit diagnosis whether rocks are in place or are transported blocks.

Ages of younger rocks are poorly known except for a single K-Ar date of 6.4 ± 0.2 Ma by Westercamp and others (1985) from igneous rock from a small excavation near Station Hill (Fig. 21). The K-Ar dates of Mayreau Basalt, 12 and 14 Ma (Fig. 21, Table 1), however, probably record the emplacement of young dikes. Moreover, the younger rocks postdate folding and foliation development in the Eocene rocks. Thus, the younger rocks are probably entirely Neogene and related to the present Lesser Antilles magmatic arc.

Microdiorite occurs in plutons and dikes. It is commonly equigranular, fine-grained, homogeneous, much-altered feldspar-chlorite rock with thin chill margins. At a few places it grades to coarser phases: medium-grained pyroxene porphyry and medium-grained quartz diorite. Microdiorite is commonly enveloped by diatremal breccia. Dikes extend from microdiorite plutons into diatremes.

Breccia comprises several types, each containing unsorted angular fragments mainly of volcanic origin. Each type is of igneous origin; none is sedimentary. The first type is autobrecciated dikes and plutons, which include monolithologic igneous fragments. The second consists of mixed igneous fragments that mainly represent exposed Neogene rock bodies. The third type occurs in proximity to coherent wall rocks and contains fragments of such wallrocks, including red carbonate and pillow basalt that are almost certainly derived from the Eocene succession. Such breccias commonly wall thin dikes and include dike tips. The third breccia type is of probable diatremal origin. Because the second type grades to the third at a few sites, the second may represent interior zones of large diatremal masses.

The undifferentiated unit (u, Fig. 21) contains dikes of uncounted sequences and wide petrographic variation. In unit N at the northern end of Mayreau (Fig. 21), as many as four sequential dikes can be seen. The principal dike parameters are phenocryst composition and dike thickness. Almost all are phenocryst-rich, from 25 to 75% phenocrysts; the phenocryst content is correlated with phenocryst size, which is as coarse as 2 cm. Dikes include plagioclase-only and clinopyroxene-only types and types with plagioclase plus clinopyroxene in all proportions. The pyroxene-rich dikes tend to be thickest (≥10 m), as exemplified by the large bodies at Tarzan and western Lande-Ci (Fig. 21). Many dikes are much altered and(or) autobrecciated. The dikes are commonly more altered than immediately adjacent wallrocks. There is no striking correlation among composition, thickness, or degree of alteration with sequence of emplacement.

The rock sampled for K-Ar (Fig. 21) by Westercamp and others (1985) is a massive hornblende-plagioclase porphyry. It

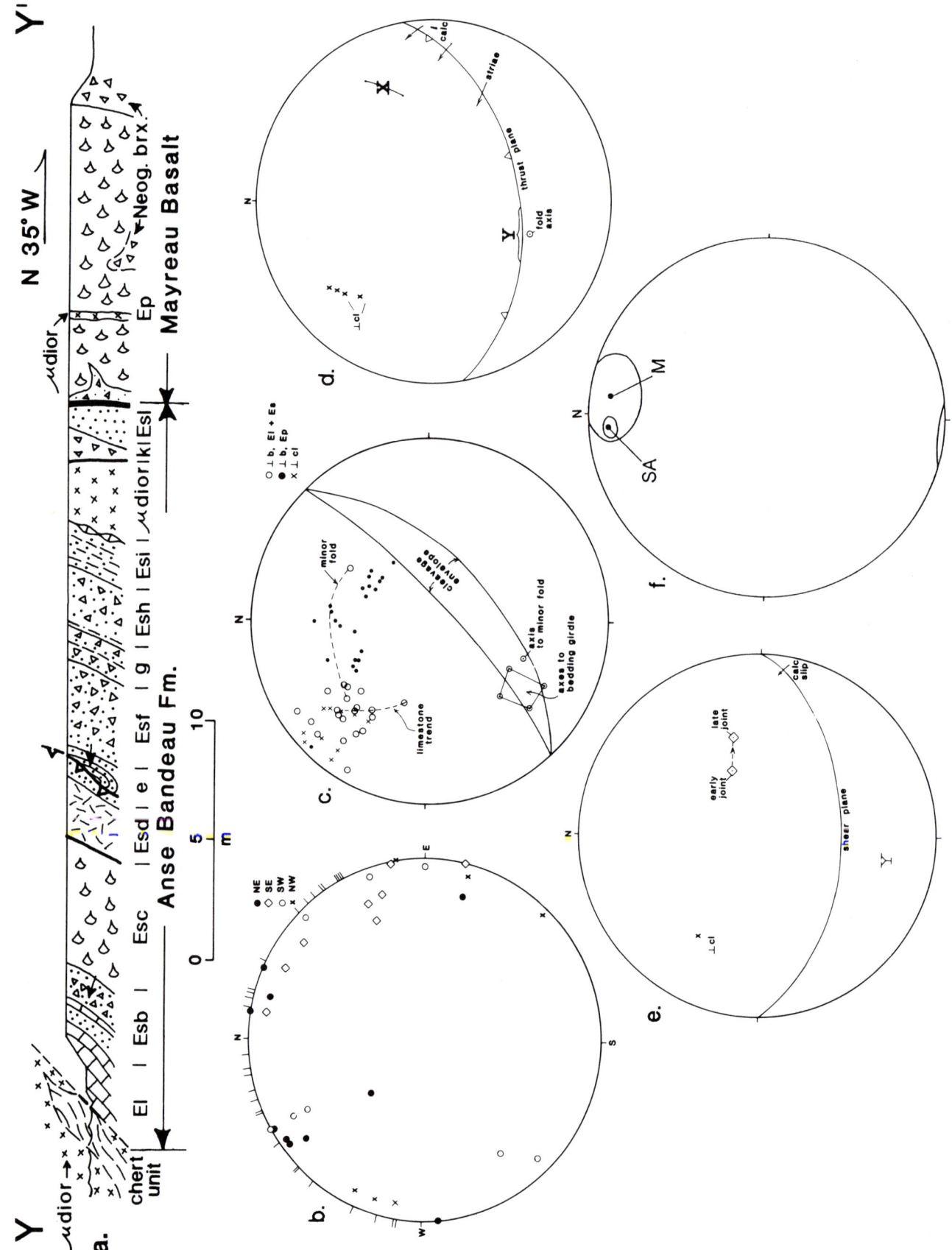

has a light-colored, dense matrix. This rock type occurs sparingly in the dike complex, and its place in the intrusion sequence is unknown.

We measured orientations of 35 dikes on Mayreau (Fig. 22b). The dikes occupy widely ranging orientations, and some individual dikes are smoothly curving with strike changes up to 90°. There is a rude preferred strike of northwest in the southeast and northwest areas of Mayreau but little or none in northeast or southwest Mayreau. The relationship of dike age to orientation is unknown. There is no evident preferred strike for the island as a whole, and Martin-Kaye's (1960) claim that the preferred strike is N60°W is incorrect.

Structure. Structures in the older rocks of Mayreau are folds, foliations, thrusts, shear bands, and the responses to intrusion of the younger rocks. The only recognized tectonic effects in the younger rocks except for those of sequential intrusion are sparse high-angle faults of uncertain sense and magnitude of slip. Owing to the absence of young sedimentary rocks on Mayreau, it is uncertain whether or not Neogene tilting has occurred.

An interpretation of the macroscopic structure of Mayreau is illustrated in the section of Figure 21. This is controlled by bedding in Eocene strata and by the form and distribution of younger intrusions. The latter are strongly conceptual because the contacts of the intrusive complexes, shown on Figure 21 to be steeply dipping, are generally poorly exposed and because the shape of such complexes is probably three dimensional and not easily extrapolated.

The Eocene succession occupies a probable train of open and(or) close folds with widths between 20 and 250 m, apparently correlated with local stratal thickness. Such folds have southwest-plunging axes (Fig. 22c) according to the bedding girdle, but the orientation of the axial planes and their asymmetry cannot be directly measured. As conceived, the train of Eocene folds is broken and extended horizontally by Neogene intrusions about 40%. The direction of principal extension is uncertain, and the distribution of dike orientations (Fig. 22b) suggests there is no maximum and that extension could be uniform horizontally.

Another macroscopic fold is implied by the bedding orientations along strike in the well-layered upper limestone of the Anse Bandeau Formation (El, Fig. 21). A short girdle indicates that open folds with east-plunging axes (Fig. 22c, limestone trend) may exist in the limestone. It is a question whether such east-plunging folds are superposed with the southwest-plunging folds that seem to pervade the entire Eocene succession or whether the limestone beds are detached from and more complexly folded than lower strata. Because the top of the limestone is a low-angle fault, it is reasonable that the limestone may include local fault-related folding. The sequencing of the two fold sets is unknown.

Spaced cleavage exists in Eocene beds and at places, in sediment between pillows. In the fine-grained basaltic clastic beds, cleavage is marked by dark discontinuous thin foliae that are probably dissolution residues. Cleavage poles occur in the same broad girdle as bedding in the Eocene units and thus, cleavage contains the southwest-plunging macroscopic fold axis (Fig. 22c).

A faulted tight class 1c to 2 minor fold exists in unit Ese of the Anse Bandeau Formation (Figs. 22a, c). The fold's downward diverging axial plane cleavage approximately parallels cleavage elsewhere in the Eocene units, and the axis of the minor fold is nearly colinear with the southwest-plunging zonal axis for all Eocene beds (Fig. 22c). The associated fault is a thrust that cuts across the intermediate limb, dips south-southeast, and has northwest-trending slip, according to striae and fault-cleavage relationships (Fig. 22d). The minor fold is the remnant of a probable asymmetric anticline-syncline pair that developed ahead of the propagating thrust tip. The layer styles and well-developed axial plane cleavage in the minor fold suggest deformation under conditions of low rheologic contrast and defluidization. Such deformation was probably controlled by elevated temperature rather than by nonlithification because sediment dikes and convolution are absent.

The general parallelism of axes of the southwest-plunging macroscopic folds and the minor fold and moreover, of both these axes with cleavage implies all three structures are kinematically related. This implies the axial planes of the major folds dip approximately steeply south-southeast. This orientation is imprecise because the cleavage represents only a small interval of the macroscopic fold train that may be locally openly refolded and because the measured cleavage could be on the limb of a cleavage fan.

South-dipping shear bands, 1 to 5 cm thick, exist in the upper biomicrite (unit El, Fig. 22e). These are identified by rotated cleavage and by successions of extension fractures within the bands. The early fractures are sigmoidal whereas the latest are planar. The shear band elements indicate left-oblique thrusting, as does the fault in unit Ese of the Anse Bandeau Formation.

The orientations of the macrofolds, the minor fold and related thrust, cleavage, and shear bands of Eocene rocks of Mayreau indicate related development of these structures in horizontal contraction with an approximately northwest-southeast trend of the principal component. The vergence of the minor fold and shear bands is northwesterly, which may reflect local failure during flexural slip on the flank of a major fold. The existence or direction of vergence of the contractile structures as a whole is unknown.

Figure 22. a, Section y–y' on Mayreau through area b) of Figure 21; b to e), structural orientation data for Mayreau; b, poles to dikes divided among four geographic quadrants; strikes of dikes also shown by ticks on exterior of net; c, poles to bedding divided by formation, poles to cleavage, and fold axes; d, data for thrust fault cutting unit Ese of Anse Bandeau Formation (Fig. 21b); X and Y are principal stretches; striae is observed slip direction, calc is calculated slip direction; e, data for shear zone in unit E1 of Anse Bandeau Formation; f, paleomagnetic data for Mayreau Basalt; dot M is site-mean direction and small circle is two-sigma uncertainty; dot SA is calculated middle Eocene direction for continental South America at Mayreau; small circle about SA direction is two-sigma uncertainty.

TABLE 4. LITHIC SUCCESSION OF ANSE BANDEAU FORMATION, SUBUNIT DESIGNATORS KEYED TO FIGURES 21 AND 22

Chert Unit of Mayreau (and microdiorite)

Chert: thin to medium bedded, light colored to deep red; several types: 1) vitreous and containing abundant (to 25%) siliceous or silicified microfossils; 2) vitreous to murky matrix, including fine clastic grains; and 3) carbonate-bearing, possibly a silicified limestone. Beds are broken and phacoidal but subparallel to fault zone. *Microdiorite* is very fine grained holocrystalline intrusion with chloritic matrix locally grading up (away from basal contact) to medium grained pyroxene porphyry; nearly envelops chert and includes red limestone xenoliths; intrudes low-angle fault as indicated by chilled margins; locally very altered to orange clayey rock.

Anse Bandeau Formation

Red biomicrite (El): plum red thinly parted limestone minorly to extensively bleached; forams exist but no siliceous microfossils are recognized; locally laminated on mm scale and some convolute lamination; bedding-parallel burrows at several horizons; hard and soft diagenetic layering; cut by weakly developed spaced cleavage and shear bands containing sigmoidal joints.

Basaltic sediments and red biomicrite (Esb): volcanigenic sediments are mainly coarse angular basalt-clast deposits, including massive, pebbly sandstone, graded plagioclase-lithic sandstone with pebbly basal zone, coarse tail grading of pebbles and distribution grading of sand matrix, and sedimentary breccia. Most fragments are very fine grained red basalt but scoria and plagioclase porphyry also present. There is no rounding. Poorly developed spaced cleavage. Red biomicrite is a tabular 20-cm-thick layer petrographically like El.

Pillow basalt (Esc): large (0.5–1 m) pillows of low ellipticity; pillows are coarsely amygdular (20% CO_3 and chlorite amygdules); pillow matrix consists of local red CO_3–SiO_2 rock with microfossils and of pebbly tuff with small pillow fragments; red veins cut pillows; underlain by 10 to 20-cm-thick fault zone with foliated fault rock and quartz veins. This is highest Eocene magmatic unit on Mayreau.

Basaltic sill (Esd): tabular zoned basalt sill; chilled top below fault contact with Esc and below that, a 1-m, strongly amygdular, zone with tubular amygdules normal to contact; below that, nonamygdular very fine grained basalt (1 m); basal 20 cm are chill zone composed of spheres of ? zeolitized rock. Sill was probably last Eocene magmatic event on Mayreau.

Basaltic turbidites (Esc): thick-bedded pebbly sandstone graded to medium-grained, plane-laminated sandstone (Tab); also several thin beds of medium-grained sandstone graded to black mudstone (Tbcde); pebbles are vesicular basalt. Unit includes a faulted tight upright syncline (Fig. 22) and strong axial plane cleavage.

Basaltic debris flow (Esf): mainly massive pebbly wacke; homogeneous outsized floating pebbles and boulders (up to 0.5 m) of red amygdular basalt with good dimensional orientation but no stratigraphic concentration or rounding; matrix is poorly sorted feldspar-lithic sandstone with cleavage; top meter is graded sandstone without pebbles.

Pebbly sandstones (Esg): two graded basaltic sandstones, each 0.75 m thick; include scattered boulders up to 0.75-m diameter, cleavage.

Basaltic debris flow (Esh): like Esf; scattered floating amygdular basalt boulders up to 0.3 m; graded top; unit is increasingly altered and veined (quartz) downward.

Basaltic sediments and limestone (Esi): basaltic sandstone with and without floating pebbles; strongly altered; red limestone at unit base at intrusive contact; limestone in phacoids.

Microdiorite: intrudes sections; undated but probably Neogene.

Altered pebbly sandstone (Esk)

Sandstone (Esl): coarse-grained feldspar-lithic sandstone, including floating basaltic granules; strong cleavage.

Mayreau Basalt (Ep)

Top of thick pillow basalt of unit Ep (Fig. 21); top of unit is a fault; includes a crosscutting intrusive pebbly sandstone in topmost zone; pillows fine upward in top 2 m; unit interior (Fig. 21) includes igneous breccia zones of 20 cm wide touching basalt fragments in dark red amygdular matrix.

Paleomagnetic data from Mayreau basalt. Paleomagnetic data were collected from the Mayreau basalt by M. Beck, R. Burmester, and R. Speed (manuscript in preparation) at nine sites, each including at least three drilled cores that yielded stable directions. Such cores have single-component magnetization and responded well to thermal and alternating frequency demagnetization. Both polarities are present. Site-mean directions cluster more tightly upon structural corrections. The magnetizations, hence, are probably primary and Eocene. Of all site-mean directions, six have acceptably small uncertainty ($\alpha \leq 15°$). The mean and two-sigma uncertainty of these six site-mean directions is plotted on Figure 22f. Figure 22f also shows a magnetic direction for continental South America at Mayreau at middle Eocene time. This direction was calculated from the middle Eocene paleomagnetic pole for North America (Diehl and others, 1983) rotated to a paleomagnetic pole for South America using interpolated stage poles of North America–Africa and South America–Africa from Engebretson and others (1985). The uncertainty fields of the two directions overlap whereas the mean direction of the Mayreau Basalt is 11° clockwise but of similar inclination to that of Eocene South America. The directions indicate that the Mayreau Basalt has undergone little or no rotation and no latitudinal movement with respect to Eocene South America.

Canouan

Canouan (Fig. 1) exposes a stratigraphic succession of older sedimentary rocks, of which a middle interval is dated middle Miocene. These older rocks are intruded and overlain by Neogene magmatic rocks (Fig. 23). The sedimentary rocks are mainly volcanigenic and clastic or pelagic carbonate types. We divide the older rocks among three units (Fig. 23), based on differences in lithic components and deformation. The Canouan Formation, a mainly clastic carbonate succession that contains the dated beds is the middle unit; it was defined by Martin-Kaye (1969) and studied further and dated by Westercamp and others (1985). We call the other units of older rocks the upper unit and lower unit of Canouan. We have attempted to date the upper and lower units without success. The lower unit and the Canouan Formation are moderately deformed whereas the upper unit and the Neogene magmatic rocks are little deformed. Our field work was concentrated in the accessible southern half of Canouan, and the northern half may include rock units different from those discussed here or by Westercamp and others (1985).

Lower unit of Canouan. The structurally lowest rocks of Canouan are undated volcanigenic-clastic carbonate turbidites, here grouped as the lower unit (Fig. 23). We infer they are unconformably overlain by the Canouan Formation. The bottom of the lower unit is not exposed. We divide the lower unit into lower and upper subunits on the basis of grain size where the strata occur in a succession north and east of Charles Bay (Fig. 23). We correlate the rocks at Kate Hill (Fig. 23) with those of the lower unit undifferentiated by rock type and deformation.

Westercamp and others (1985) interpreted the rocks of the lower unit as pyroclastic breccia and tuff and magmatic intrusions and called the unit an ancient volcanic massif. In contrast, we regard the volcanigenic deposits as resedimented, not necessarily from a local vent, and the intrusions as unrelated and belonging to the Neogene magmatic complex.

The lower lower unit is a succession mainly of coarse-grained sandstone and sedimentary breccia with subordinate calcarenite that occur in thick graded beds. Clasts are chiefly volcanilithic. The upper lower unit is volcanigenic sandstone and interbedded black volcanic calcarenite. The lithoclasts are plagioclase-rich volcanic rocks, some of which are porphyritic, together with very fine grained siliceous lithics, discrete plagioclase, and phyllosilicate. Postdepositional epidote and micaceous grains indicate low-grade metamorphism. Intraclasts of mudstone and sandstone occur. The carbonate particles are mostly recrystallized such that their precursors are uncertain. Some, however, are clearly skeletal. The beds are tabular and graded and include bottom current features that indicate deposition by turbidity currents below subwavebase.

In the muddy-sandy turbidites of the lower unit, there is a weak microscopic conjugate foliation of recrystallized phyllosilicates. The foliations are at right angles and diverge around clastic grains.

Rocks along the east coast south of Riley Bay (Fig. 23) are fine-grained volcanigenic and calcarenitic turbidites and include at one horizon, red pelagic mudstone. These beds are probably continuous with those of the lower unit at Charles Bay, and we interpret them to be in the upper lower subunit (Fig. 23, A–A′).

The sedimentary rocks of the lower unit undifferentiated at Kate Hill (Fig. 23) are tabular muddy turbidites in a fining and thinning upward sequence a few tens of meters thick. The sand particles include volcanic lithoclasts, plagioclase, quartz, chert and (or) siliceous mudstone, and skeletal types. At least one specimen contains microscopic conjugate foliation, and the beds are strongly folded (Fig. 23C) compared to those of the Canouan Formation and the upper unit. Such characteristics are the basis for our assignment of the Kate Hill beds to the lower unit, which is contrary to their assignment by Westercamp and others (1985) to an upper volcanic system younger than the Canouan Formation.

Canouan Formation. The Canouan Formation is a well-bedded succession, about 100 m thick, of black carbonate-rich turbidites. The beds are mainly fine-grained calcarenite and micrite. The carbonate particles are benthic and planktic foraminifers, red algae, and other skeletal components. Volcanigenic clastics, like that of the lower unit, occur abundantly in some beds but only sparsely in others. The muddy zones of beds are thoroughly burrowed.

Westercamp and others (1985) identified nine species of planktic foraminifers from beds near Quarry Point, apparently over a 10-m-thick section (Table 2; Fig. 23B). These give a composite age of N5, *P. glomerosa,* which is early middle Miocene, about 16 ± 0.5 Ma. Benthic forams identified in this interval by Westercamp and others (1985; Table 2) are compatible with zone N5. Because the Canouan Formation is turbiditic, N5 can only be regarded as a maximum age. The temporal coexistence in a single zone of all of a large number of planktic and benthic species, however, implies that resedimentation must have been penecontemporaneous with initial deposition. We therefore regard at least an interval of the Canouan Formation as early middle Miocene.

The Canouan Formation is structurally higher than the lower unit. The contact between them, however, is not exposed and could be either an unconformity or a fault. We interpret it as an unconformity because there is no fault-related rock in scree covering the contact and because the Canouan Formation would be a compatible continuation of the upward fining and the upward carbonate enrichment of the lower unit.

Upper unit of Canouan. We designate as the upper unit (Fig. 23) pebbly marl and clastic carbonate that is structurally above the Canouan Formation. It is poorly bedded and not turbiditic. The pebbles and boulders are mainly basalt and occasionally, black limestone like that in beds of older units. The upper unit also includes minor thin basaltic sandstone and dikes of basalt whose time of emplacement is unknown. The clastic carbonate particles include red algae and broken benthic foram tests. Aside from such lithic differences from the Canouan Formation, the upper unit also is distinguished by its subhorizontality from the folded Canouan Formation, with which it was included by Westercamp and others (1985). There are no dates of the upper unit, but it must be middle or late Miocene by stratigraphic relations.

Neogene magmatic rocks. The magmatic rocks of southern Canouan are either intrusive or protrusive with respect to the

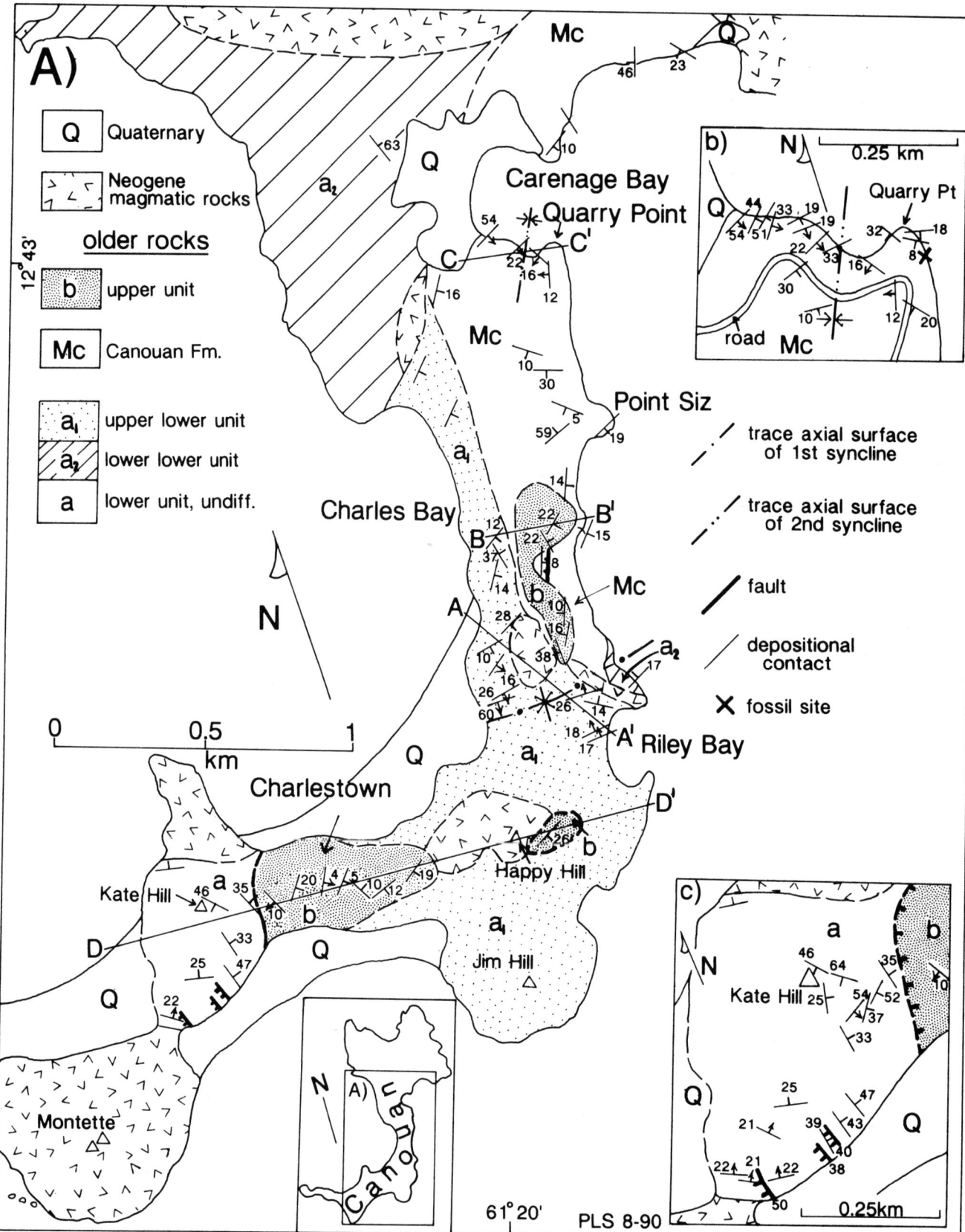

Figure 23. A, Geologic map of part of Canouan by P. L. Smith; b, enlarged map of Quarry Point area, showing site of dated fossils in Canouan Formation; c, enlarged map of Kate Hill area.

sedimentary units, discussed above. Westercamp and others (1985) obtained an age of 6.8 Ma by K-Ar (whole rock) for a basalt dike within the magmatic complex of Montette (Fig. 23). Therefore, the onset of Neogene magmatism in Canouan occurred in Miocene time between 16 and 6.8 Ma.

Structures. The principal structures of Canouan are macroscopic folds of parts of the lower unit and the Canouan Formation and normal faults that are probably younger than the folds.

The upper lower unit contains an open macroscopic syncline with east-northeast axial trace near Riley Bay (Figs. 23, 24, section A–A′). The limb attitudes are indicated by clusters of bedding orientations (Fig. 25a), and the fold's zonal axis plunges shallowly east-northeast. Together with the bisecting trace between the limbs, the axis defines an approximately vertical axial plane that strikes N70°E.

Most of the Kate Hill beds of the lower unit are partially girdled about a zonal axis (axis 2, Fig. 25a) that is close to that of the Riley Bay syncline. The near-coaxiality is one of the reasons we correlate the Kate Hill beds with the lower unit. The array of bedding attitudes at Kate Hill, however, does not define a single macroscopic axial trace, but rather, is suggestive of high frequency folds or a faulted fold.

Beds of the Canouan Formation at the south shore of Carenage Bay (Fig. 23) are folded in a major syncline with axial trace trending N24°E. Such bedding is girdled about a zonal axis that plunges shallowly south-southwest (Fig. 25b), and all bedding elsewhere in the Canouan Formation is approximately cylindrical to that axis (Fig. 25b). Therefore, we interpret the Canouan Formation as a whole to be folded with a north-northeast-striking, subvertical axial plane.

Bedding in the upper unit has dips that are all shallow and without preferred direction (Fig. 25c).

Three normal faults in the lower unit at Kate Hill dip moderately southwest (Fig. 25a). These probably record postfolding extension and may have accompanied Neogene magmatism. We infer an east-dipping normal fault below cover east of Kate Hill to account for the juxtaposition of the lower and upper units. We also interpret a normal fault klippe of the upper unit above the lower unit at Happy Hill; the fault is inferred because bedding in the hangingwall is more steeply dipping than the contact. A fault in the Canouan Formation dips moderately west-northwest (Fig. 25b).

Structural interpretation. The structures of the three sedimentary rock units differ significantly: folds with east-northeast-trending axial trace in the lower unit; an open macroscopic fold with north-northeast-trending axial trace in the Canouan Formation; and a subhorizontal stack in the upper unit. Assuming we are correct in interpreting the three units to be a depositional sequence, the history of deformation is as follows. First, folding affected the lower unit before or during the development of the basal unconformity of the Canouan Formation in early middle Miocene time. Second, folding with north-northeast axial trace reoriented the folds of the lower unit and deformed the Canouan Formation during middle or late Miocene time before deposition of the upper unit. Post–upper unit deformation is slight tilting in various directions, perhaps related to normal faulting and to magma emplacement, which occurred between middle Miocene and the present.

As a test of the idea that the folds of the lower unit have been superposed by the syncline of the Canouan Formation, we

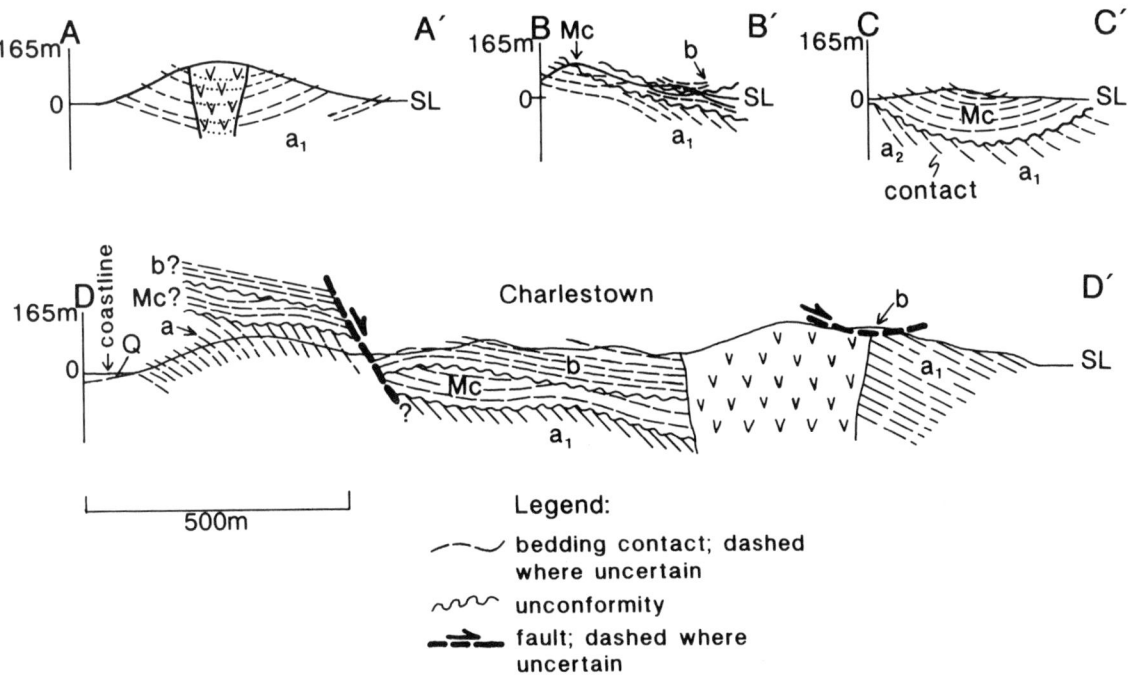

Figure 24. Sections of Canouan, located on Figure 23.

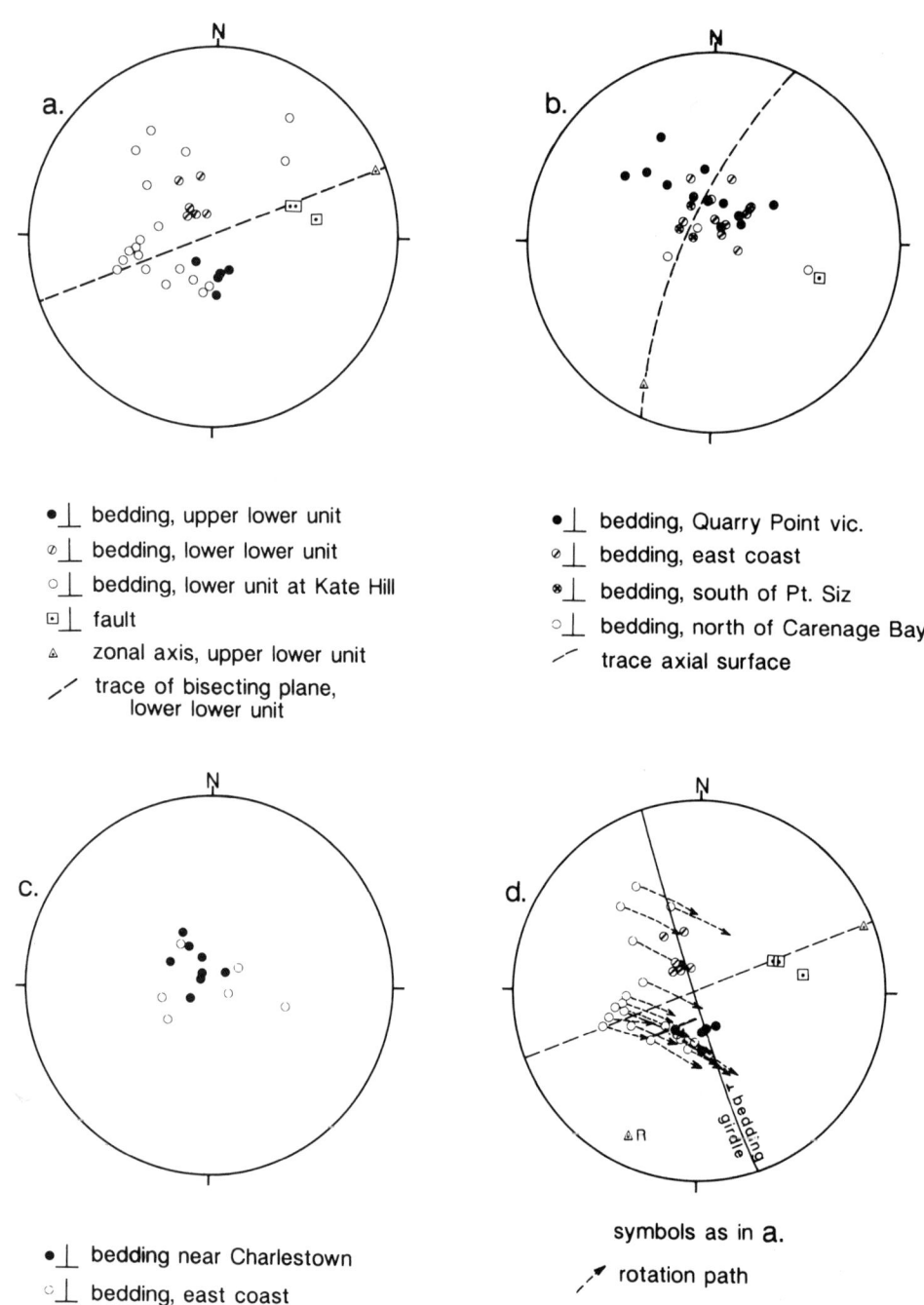

Figure 25. Orientation data for Canouan: a, lower unit; b, Canouan Formation; c, upper unit; d, Rotation of poles to bedding in lower unit at Kate Hill (Fig. 23C) about axis R.

rotated the beds of the lower unit west of the syncline's projected axial trace (Kate Hill beds) about the syncline's axis (Fig. 25d). A rotation of 30° brings the Kate Hill beds into close juxtaposition with the partial girdle of beds of the lower unit east of the synclinal axial trace. 30° is approximately average dip of the western limb of the syncline (Fig. 25b). The rotation is consistent with the hypothesis.

We interpret the successive folds of Canouan to be products of horizontal contraction, the earlier phase with approximately NNW-SSE bearing and the second WNW-ESE bearing. An alternative, that one or both fold phases could be of fault-accommodation origin and not necessarily contractional, is conceivable but not adopted here because no corresponding faults can be identified (e.g., at the base of the Canouan Formation).

Evolution of Canouan. The earliest record of Canouan is the deposition of the lower unit in early Miocene or earlier time. Below, we suggest the age is early Miocene because of similarities of the lower unit and the Canouan Formation. The site of deposition was basinal and below wavebase. The sediment of the lower unit was transported by turbidity currents and probably, debris flow from active arc volcanoes that included fringing carbonate cover. The lower unit records a general upward fining and thinning and increasing clastic carbonate content. This suggests either waning of magma discharge or progressive bypassing of the basin by the principal sediment conduits from vent regions during deposition of the lower unit.

This was followed by folding with approximately NNW-SSE horizontal contraction. It is probable that such folding was concurrent with the development of a suprajacent submarine erosion surface that became the unconformity at the base of the Canouan Formation. It is unclear whether the microscopic foliations and metamorphism of the lower unit developed during first or later deformation.

The Canouan Formation was laid down on the moderately deformed lower unit over an interval that includes early middle Miocene time (about 16 Ma). The depositional system of the Canouan Formation was similar to that of the lower unit but further evolved along the trend of upward fining and clastic carbonate enrichment. The Canouan Formation may record a period of volcanic quiesence or major bypassing of its basin by contemporaneously erupted debris. The lithic trend between the lower unit and the Canouan Formation suggests the two units accumulated in near-continuity, separated by a brief phase of contraction that did not destroy the basin. Thus, we infer that the lower unit is close to the Canouan in age, perhaps early Miocene or late Oligocene rather than a more ancient deposit. The WNW-ESE contraction of the Canouan Formation and underlying rocks occurred in middle or late Miocene time before the deposition of the upper unit.

The upper unit was deposited in basinal conditions that permitted the accumulation of both pelagic and clastic carbonate and minor basaltic debris on a substrate of folded Canouan Formation. Such deposition occurred in middle or late Miocene time, before the advent of Neogene magmatism, known on Canouan only to be 6.8 Ma or older. It is uncertain whether the upper unit accumulated in the same basin as the Canouan Formation or in a newly established one, following the second phase of deformation. If it was the same basin, the upper unit may represent a cessation of the sediment-gravity flows that delivered the earlier sediments and perhaps a shallowing to and oscillation of seabed above and below neritic depths.

Mustique

Mustique is the northernmost of Grenadine islands examined for this report. It exposes a unit of older rocks, Neogene(?) magmatic rocks, and a third unit of undifferentiated sediments and rock (u, Fig. 26) that encompasses unexposed and unstudied tracts. The unit of older rock is a probably homoclinal sequence of volcanigenic sedimentary rocks, part of which were dated Oligocene by Westercamp and others (1985). This sequence was incorrectly called metamorphosed volcanic rocks and tuff by Martin-Kaye (1969) and hyaloclastites and lava by Westercamp and others (1985). We argue below that the unit of older rock of Mustique contains no contemporaneous magmatic rock and that igneous rocks previously claimed to be lava in the section are in fact intrusive.

Older rocks of Mustique. This unit consists entirely of homoclinal south-dipping volcanigenic sedimentary rocks (Fig. 26). The homocline spans all of Mustique from north to south. If there are no unrecognized faults that repeat or omit section, the section is at least 2 km thick. Sedimentary structures indicating facing were found only near Gun Hill (Fig. 26), and this implies but does not prove the whole section is upright.

The older rocks were sampled by Westercamp and others (1985) from beds near Black Sand Bay and Gallicaux Bay (Fig. 26), which probably are in the same several hundred meter thick stratigraphic interval high in the exposed section of the unit of older rocks. They identified planktic foraminifers (Table 2) and nannoflora *(Sphenolithus distensus, Cyclicargolithus abisectus, Dictyococcites dictyodus, Helicosphaera euphratis,* and *H. perchnielsenae).* The composite age range is within P19 or upper NP23 to NP24, late early to early late Oligocene. Lower beds of the unit could be substantially older.

The volcanigenic sedimentary rocks are characteristically coarse grained and thickly bedded. They consist mainly of lithic grains from medium sand to boulders as coarse as 1 m, together with feldspar, clinopyroxene, and amphibole sand. Rock types are massive sandstone, plane and dune-laminated sandstone, pebbly sandstone, and sedimentary breccia. All these are commonly tabular and poorly sorted and stratified but have good preferred orientation of nonequant pebbles. The coarse sediments are mainly nongraded except for a few layers with upward diminishing pebble/sand ratios.

Finer-grained sediments occur in a few intervals, most thickly near Black Sand Bay (Fig. 26). There, muddy and sandy thin beds alternate without grading. Mudstones are calcareous and sandy and include feldspar and lithic sand, forams, and radiolarians. Sandstones are commonly plane-laminated and finely pebbly. This interval also contains channelized beds that fine upward from microbreccia to marl. The graded beds occur in repeated sets. Channels are as wide as 3 m and are up to 1 m deep.

Volcanigenic particles of the older rocks are almost entirely angular; slight rounding occurs in only a couple of beds. The clasts are commonly polymict but related arc-derived volcanic types, chiefly feldspar or pyroxene porphyry, feldspar-pyroxene porphyry, scoria, and massive dark microporphyry and aphanite. Carbonate clasts are concentrated in a few beds; these include boulders of biogenic rock, either coralline or algal, and sand-sized benthic forams.

Debris in the older unit is certainly mainly first cycle volcanigenic debris from an arc-type source. The occurrence of carbonate fragments of probable shallow-marine origin in some beds indicates a shoaled source region. The prevailing angularity indicates that residence of debris at the source was brief and in fact, much may have entered the sea as pyroclastic flows. The sediment transport, however, was mainly water-rich and fully turbulent. The site of deposition of the upper quarter of the succession was basinal, as indicated by marls bearing pelagic fauna. The lower section was either also basinal or a canyonfill, as interpreted by the paucity of rounding that would be expected if these coarse, high-regime deposits had evolved in a littoral environment. The direction of the sediment source relative to Mustique is uncertain.

Tectonic structures in the sedimentary rocks are joints in calcareous thin beds at Black Sand Bay and the shallow homoclinal tilt. In particular, cleavage and mesoscopic folds are absent.

Igneous intrusions in older unit. Westercamp and others (1985) stated that flows of andesitic lava are interbedded with the sedimentary rocks of our older unit at Pasture and Gallicaux Bays (Gallicaux is Lagoon Bay on their map). We find such igneous rocks are intrusive and postsedimentary, as indicated by contact relations and by K-Ar dates.

At Pasture Bay, the principal intrusions are porphyritic (Fig. 26B and C); they occur in several roughly tabular bodies that dip more shallowly than and have irregularly configured contacts with the sedimentary rocks. In general, the contacts have a staircase geometry where they alternately cut across and parallel sedimentary layering. Hence, an igneous body alternates downdip as a dike and sill; the dikes tend to cut coarser (pebbly) layers whereas the sills tend to occur in sandier layers. Further, the boundaries of the igneous bodies are locally irregular on the 1-m scale. For example, small pods of igneous rock intrude sediments

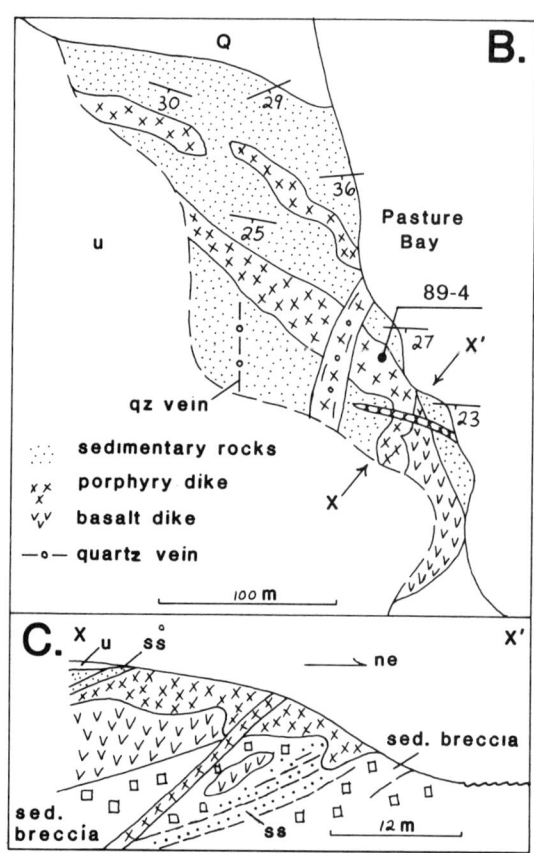

Figure 26. A, Geologic map of Mustique; B, map of area at Pasture Bay showing succession of dikes; C, section x–x', located on B.

from the tops of the igneous bodies. The igneous bodies show evidence of multiple intrusion in the form of wedges and lenses of brecciated aphanitic igneous rocks along the walls of the intrusions; such breccias are intruded by the internal masses of porphyry.

The intrusive sequence at Pasture Bay is complex (Fig. 26B, C). Early intrusions include both porphyry and basalt in uncertain time relation. Late intrusions are two nearly orthogonal thin dikes of porphyry. One of these is highly altered and contains abundant quartz veins whereas the other is largely fresh.

At Gallicaux Bay (Fig. 26), bodies called lavas of andesite porphyry by Westercamp and others (1985) are in fact dikes of basalt that is petrographically like basalt of the Neogene(?) unit. The northernmost dike at Gallicaux Bay is thick, more than 3 m, whereas those to the south are in an irregular network of thin bodies.

We radiometrically dated two dikes that invade the older rocks of Mustique to test our observation that such bodies are younger than the Oligocene host rocks whose ages are between 28 and 33 Ma or older. A dike at Gallicaux Bay (89-6, Fig. 26A), a fresh plagioclase-clinopyroxene holocrystalline basalt, gave a K-Ar whole-rock date of 19.2 ± 0.6 Ma (Table 1). The intruded strata at this site are on strike with the beds dated by fossils near Black Sand Bay (Fig. 26A). The date thus supplies further evidence that the basalt is intrusive, not an extrusion.

A second dated dike is the early porphyry at Pasture Bay (89-4; Fig. 26B). We dated whole-rock samples of fresh plagioclase porphyry by K-Ar (25.2 ± 0.4 Ma) and $^{40}Ar/^{39}Ar$ (Table 1). The $^{40}Ar/^{39}Ar$ total gas age is 25.7 Ma, and the spectrum indicates a near-plateau of about 25 Ma between 10 and 95% of gas released (Fig. 5g). We also dated by $^{40}Ar/^{39}Ar$ phenocrystic plagioclase from 89-4 (Fig. 5f). This gives a total gas age of 25.2 Ma. The spectrum of the plagioclase, however, is very discordant; it has a central subplateau near 25 Ma but much younger dates in the first 25% and older dates, exceeding 35 Ma, in the last 25% of gas release. The origin of discordances in the plagioclase spectrum is uncertain. We assume from the whole-rock dates on 89-4 that its age is approximately 25 Ma and that the igneous body is younger than its wallrocks, which are stratigraphically about 100 m below the dated Oligocene horizon.

Younger rocks. This unit contains intrusive and possibly extrusive rocks that are undated but younger than the Oligocene strata and probably of Neogene age. These include basalt in the northern quarter of Mustique and hornblende-quartz diorite west of Macaroni Bay (Fig. 26). Plagioclase and clinopyroxene porphyry dikes and sills that cut the older rocks may also belong to the suite of younger rocks.

SYNTHESIS OF ISLAND GEOLOGY

Below, we synthesize the geology of the older rocks of the southern Lesser Antilles arc platform in four topics: stratigraphy, magmatism, depositional environments, and deformation. Table 5 summarizes the properties of each of the older units discussed in the previous section. Figure 27 displays age-range data: zonal, permissible, composite, and stratigraphic, as defined earlier, for the older units; the time scale and fossil zone correlations used are those of Berggren and others (1985).

Stratigraphy

Contacts. Among the 26 units of older rock in the SLAAP (Table 5), contacts of demonstrably depositional origin are few. On Carriacou, the Belvedere-Belmont-Kendeace-Carriacou-Grand Bay Formations are certainly in depositional succession as are the two sedimentary rock units on Prune Island. All other possible contacts are covered or are faults. Of these, there is evidence that depositional continuity exists or did exist on Mayreau between the Mayreau Basalt and Anse Bandeau Formation; on Carriacou, among Cherry Hill Basalt, Bogles Limestone, and Belvedere Formation; and on Canouan, the lower unit, Canouan Formation, and the upper unit. On Union Island, the massive porphyry probably intruded the sandstone-chert unit.

Unconformities. The only recognized unconformity that could represent a substantial hiatus or lacuna is that at the base of the Belmont Formation in Carriacou. It could occupy a part or the whole of the duration between nannoflora zones NP25 to NN3, middle late Oligocene to early middle Miocene, 28 to 17.5 Ma (Fig. 27). With assumptions that Belmont B beds are autochthonous and near the base of the Belmont Formation (model 1, Fig. 14) and that the subjacent Belvedere Formation was substantially eroded, the sub-Belmont unconformity probably has a short duration in early Miocene time, perhaps between 21 and 23 Ma (Fig. 27). The uncertainty is mainly in age of earliest Belmont strata.

The erosional unconformity at the base of the Kendeace Formation is probably also early Miocene, possibly early middle Miocene, given the assumptions of the prior paragraph. Its duration was probably short.

Within the Belvedere Formation of Carriacou, a gap in zonal age ranges exists in the early Oligocene (nannoflora zones NP21 and 22; Fig. 27). The gap is not correlated with conspicuous rock stratigraphic changes in the formation. We are uncertain whether the gap represents an intraformational lacuna or condensed section or a structural omission.

Continuity. A basic question is whether the older rocks as a whole evolved as a vertical succession and in close proximity to one another (within an area, say, 200 km on a side) or whether units have been tectonically juxtaposed and transported large distances relative to one another. The possibility of large relative tectonic transport certainly exists, given the uncertainty of many contacts among older units and the existence of deformation and thrust faults of probably significant displacement (Bogles thrust, Baradel thrust). In this case, correlations are invalid, and a comprehensive paleogeography cannot be inferred. On the other hand, certain features suggest the older rocks could have formed contiguously: a modest coincidence of lithic character with age and possibly related times and kinematics of deformation. Although there is no definitive answer to the question, we proceed

TABLE 5. SUMMARY OF PROPERTIES OF OLDER ROCK UNITS OF THE GRENADINES AND GRENADA

Island Unit	Zonal Age Ranges	Stratigraphic Age Range	Contacts	Rocks and Depositional Environment	Structures
Carriacou Cherry Hill Basalt	Lower NP16	Lower NP16 (middle Eocene) and below - no lower limit	Lower contact probably Bogles thrust; upper is conformable depositional contact with Bogles Limestone, and probably, Belvedere Formation	Pillow basalt of possible spreading origin; ≥ 300 m thick; interstitial pelagic CO_3 and hyaloclastite	In Bogles allochthon and in major anticline overturned to NW; local cleavage in interstitial sediments
Bogles Limestone	Lower NP16	NP16 (middle Eocene)	Concordant to Cherry Hill Basalt; either unconformably below or a diagenetic facies of basal Belvedere Formation	Limestone, chert, and cemented chalk, 10 to 30 m thick; pelagic carbonate protolith	In Bogles allochthon and in major anticline overturned to NW; local spaced cleavage
Belvedere Formation	NP15-20 and NP23-25	Lower NP16 (middle Eocene)-NN4 (early Miocene)	Probably either unconformable or conformable above Cherry Hill Basalt and Bogles Limestone; unconforable below Belmont Formation	Interstratified pelagic beds (marl, chalk, mudstone), arc-volcanigenic turbidite, and foram microconglomerate; probably >500 m thick; minor basalt near top; basinal environment with turbidite fans, mainly outerfan facies	In Bogles allochthon and in early major anticline overturned to NW; late major and minor superposed folds with N to NW axial trace
Anse La Roche Formation	NP19-21 (incl. composite)	No older bound; older than mid-NN4	Top is Bogles thrust and overlap by Belmont Formation; bottom contact not exposed	Coarse arc-volcanigenic and skeletal sediment-gravity flows and minor hemipelagite; >250 m thick and probably much thicker, base of slope channel system; inner and mid-fan facies	Early train of upright folds with ENE axial traces; later local superposed deformation related to Bogles thrust; late local spaced cleavage
Belmont Formation Unit B	NN2-3	No older bound; pre-5 Ma	No base exposed; cut by late Neogene intrusions	Coarse arc-volcanigenic turbidites from arc and basaltic source regions; >275 m thick; probably submarine fan	Train of open folds with northerly axial traces
Belmont Formation Unit A	None	NP25 or younger; mid-NN5 or older	Base is unconformity above Belvedere and Anse La Roche Formations; top is unconformity below Kendeace Formation	Poorly bedded arc-volcanigenic conglomerate and sandstone; 100 to 200 m thick; probably inner submarine fan channel system	Same as Kendeace and Carriacou Formations

TABLE 5. SUMMARY OF PROPERTIES OF OLDER ROCK UNITS OF THE GRENADINES AND GRENADA (continued)

Island Unit	Zonal Age Ranges	Stratigraphic Age Range	Contacts	Rocks and Depositional Environment	Structures
Kendeace Formation	NN3-5 (permissible age range)	NP25 or younger; mid-NN5 or older	Base is unconformity above Belmont A; top is conformity with Carriacou Formation	Arc-volcanigenic and carbonate sandstone, conglomerate, and mudstone; ≤40 m thick; possible shallow marine clastic complex	Shallowly east-dipping homocline plus several singular open folds with northerly axial traces
Carriacou Formation	NN4-5	NN3-5	Conformable bottom with Kendeace Formation; conformable top with Grand Bay Formation	Platformal clastic carbonate strata plus minor volcanigenic sandstone and mudstone; 80 to 100 m thick	Shallowly east dipping homocline plus several singular open folds with northerly axial trends
Grand Bay Formation	NN5-7	Base is NN4 or younger; top is pre-11 Ma	Conformable bottom with Carriacou Formation; no overlying strata; presumed to precede late Neogene intrusions	Mainly arc-volcanigenic sandstone in probable fluvial-littoral complex	Similar to Carriacou Formation
Union Island Sandstone-chert unit	NP 14-18 (permissible age)	Pre-12 Ma	Lower contact unexposed; probably intruded by massive porphyry unit and by late Neogene (≤12 Ma) dikes	Variably chertified pelagic rocks (marl, calcwacke) and fine- and coarse-grained, arc-volcanigenic turbidite; general basinal environment	Folds of bedding with probable S vergence, due to NS contraction
Massive porphyry unit	None	Pre 12 Ma; NP14 or younger	Faulted, probable intrusive boundary with sandstone-chert unit	Altered felsic protusion or shallow intrusion	Local foliation
Breccia unit	None	None	All contacts covered by Quaternary or arc volcanics	Volcanigenic sedimentary breccia environment uncertain	Local foliation
Prune Island Volcanigenic sedimentary rocks	None	Pre-Holocene	Unit lies above non-volcanigenic sedimentary rocks; lies below Holocene sediments	Coarse-grained, mainly basaltic turbidite plus minor pelagic; basinal environment	*Western Prune:* Train of major close folds with S vergence, ENE axial traces, and 0 to 20° E plunging axes; *Eastern Prune:* Tight major folds with S vergence, NNE axial traces, and 25° WSW plunging axes. S vergent thrust both areas; possibly early NW-SE shortening + SE overriding and later rotation about N axial trace; times unknown

TABLE 5. SUMMARY OF PROPERTIES OF OLDER ROCK UNITS OF THE GRENADINES AND GRENADA (continued)

Island Unit	Zonal Age Ranges	Stratigraphic Age Range	Contacts	Rocks and Depositional Environment	Structures
Nonvolcanigenic sedimentary rocks	None	Pre-Holocene	Lower contact unexposed; underlies volcanigenic sedimentary rock unit	Coarse- and fine-grained turbidites; grains are sandstone, chert, and mudclasts; chertified pelagic beds; basinal environment	Same as above
Jamesby Volcanigenic sedimentary rocks	Upper NP15-NP16	Pre-Holocene	Lower contact not exposed; upper contact is unconformity with Quaternary sediments	Coarse and fine arc-volcanigenic sediment-gravity flows; minor pelagic carbonate; >75 m thick basinal environment	Major synclinal fold of beds overturned to SSE; early local layer-parallel foliation
Baradel Chert unit	NP15	Pre-Holocene	Lower contact is Baradel thrust; upper contact is Quaternary sediment	Chertified calcareous pelagic beds and thin turbidites >30 m thick; distal basinal environment	Hangingwall of Baradel thrust with ESE transport; early low-angle foliation; later folds related to thrusting
Volcanigenic sedimentary rock unit	None	Pre-Holocene	Lower contact is unexposed; upper contact is Quaternary sediment; cut by ?Neogene dikes	Massive, very coarse arc-volcanigenic strata; minor sandstone; >50 m thick; basalt dikes and diatreme	Footwall of Baradel thrust; NW-dipping homocline deformed in thrust zone
Mayreau Mayreau Basalt	None	Top is upper NP15 or older; no lower bound	Lower contact unexposed; upper contact is probably Anse Bandeau Formation	Pillow basalt, pillow breccia, interstitial red pelagic limestone; >500 m thick spreading origin of basalt	Train of open or close folds with probable SSE dipping axial plans; local spaced cleavage brecciated around Neogene intrusions
Anse Bandeau Formation	NP15	Pre-Neogene intrusions, post-Mayreau Basalt	Lower faulted contact with Mayreau Basalt; upper contact is fault	Red pelagic limestone, pillow basalt, basaltic debris flow and turbidite, >35 m thick; open marine basin	Folds with SSE-dipping axial planes; internal thrust with NW slip; local fold with E-plunging axis near top, possibly related to capping fault; shear bands
Chert unit	None	Pre-Neogene intrusion	Lower contact is low-angle fault; intruded by Neogene pluton	Chertified calcareous and siliceous pelagic and fine-grained siliciclastic beds; open-marine basin	In hangingwall of low-angle fault; broken formation

TABLE 5. SUMMARY OF PROPERTIES OF OLDER ROCK UNITS OF THE GRENADINES AND GRENADA (continued)

Island Unit	Zonal Age Ranges	Stratigraphic Age Range	Contacts	Rocks and Depositional Environment	Structures
Mustique					
Older rocks unit	NP23-24	Pre-Neogene intrusion	Lower contact unexposed; upper contact with Quaternary; intruded by Neogene rocks	Arc-volcanigenic sandstone and sedimentary breccia; minor pelagic beds; base of slope and/or submarine canyon fill	Homocline, south dipping
Grenada					
Tufton Hall Formation	NP19-21	Pre-22 Ma	Lower contact unexposed; upper contact with Neogene volcanics	Arc-volcanigenic turbidite and pelagic marl, outerfan-basin plain facies	Early N-verging recumbent folds and thrusts; bed-parallel cleavage; later NS extension
Tempe-Parnassus beds	NP23-24	Pre-22 Ma	Lower contact unexposed; upper contact with Neogene volcanics	Volcanigenic-carbonate grain flow	Macroscopic anticline with EW axis
Canouan					
Lower unit	None	Pre-Canouan Formation; no base	Lower contact unrecognized; probable unconformable below Canouan Formation	Volcanigenic-clastic carbonate grain flows and turbidite; upward fining and CO_3 enriching	Early ENE-trending folds; microscopic foliation; late refolding
Canouan Formation	NN5	Pre-6.8 Ma	Probably unconformable on lower lower unit; conformable below lower lower unit	Carbonate-rich turbidite; 100 m thick	Open syncline with NNE axial trace
Upper unit	None	N5 or younger; pre-6.8 Ma	Conformable on Canouan Formation; intruded by Neogene dikes	Basinal carbonate plus minor basaltic sediment	Homoclinal

assuming the older rocks evolved as a succession over a finite region because the other case is completely unconstrained. Figure 28 presents a model time-space array of older rock units that assumes they evolved in continuity in a single basin.

Duration. The zonal age ranges of the older rocks as a whole are early middle Eocene to middle middle Miocene, approximately 50 to 13 Ma (Figs. 27, 28). The oldest zonal age (upper P10 to P11) is likely to be the actual maximum age of rocks of the SLAAP, even though we have only dated the tops of the lowest units, Mayreau Basalt and Cherry Hill Basalt. As explained below, the two basalt formations are interpreted to have originated by sea-floor spreading. The basalt in each unit is thus vertically penecontemporaneous. The minimum zonal age, however, may not be the youngest actual age because the stratigraphic bound on the youngest formation, the Grand Bay, is about 11 Ma.

Rocks earlier called Mesozoic (sandstone-chert unit of Union Island) by Westercamp and others (1985) have been reexamined and found to be Eocene.

Lithic divisions. Most of the older rocks units can be assigned with moderate certainty to one of three groups by lithic character (Fig. 28). The tightly dated units within each group indicate the groups are sequential with time. The groups are (units with asterisks are undated):

I. *Pillow basalt and pelagic sediments.* Mayreau Basalt, Anse Bandeau Formation, chert of Mayreau*; Baradel chert unit; Cherry Hill Basalt, Bogles Limestone; middle Eocene (NP15 to lower NP16; upper P10 to lower P12).

II. *Arc-volcanigenic sediment-gravity flows and pelagic sediments.* Belvedere Formation, Anse La Roche Formation, Belmont Formation (Carriacou); sandstone-chert and breccia* units of Union Island; both units* of Prune Island; rocks of Jamesby; volcanigenic sedimentary unit* of Baradel; older rocks unit of Mustique; lower unit of Canouan, Canouan Formation, and

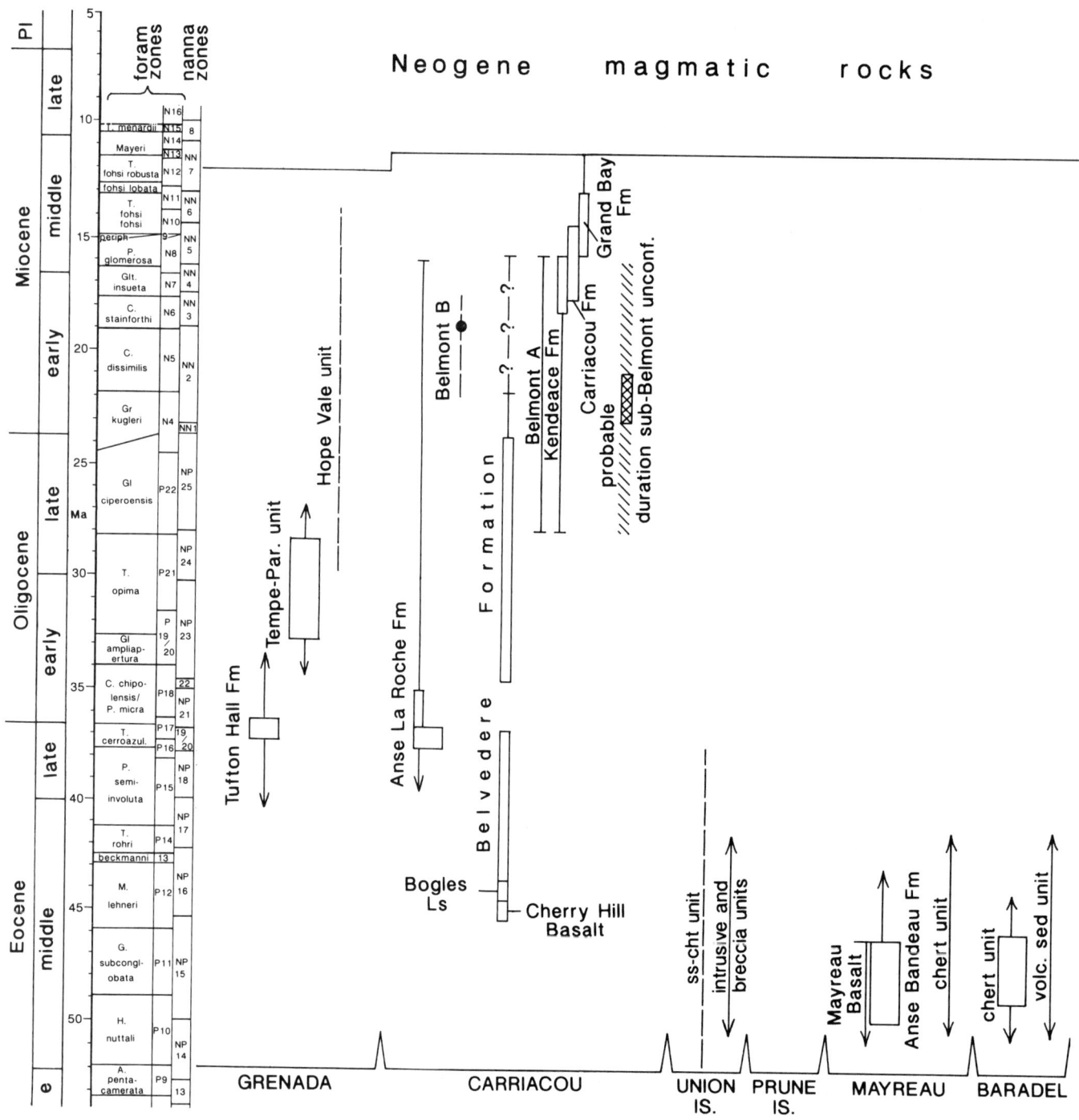

Figure 27 (continued on left column of facing page). Ages of older and Neogene rocks of the SLAAP; age nomenclature explained in text; time scale and zonal correlations from Berggren and others (1985).

upper unit of Canouan; Tufton Hall Formation and Tempe-Parnassus beds of Grenada; late middle Eocene to middle Miocene (upper NP16 to NN7; upper P12 to N12).

III. *Platformal units of Miocene age.* Kendeace, Carriacou, and Grand Bay Formations of Carriacou; probably late early to middle Miocene (NN4 to 7; N7 to 14).

The main uncertainty in this classification is the assignment of undated sediment-gravity flows in group II. The similarity of lithic character, sedimentary structures, and deformation of the Prune Island units to the Jamesby unit is the basis for assignment of Prune units. The Baradel volcanigenic sedimentary unit is not readily correlated with other units, but its inclusion in II seems

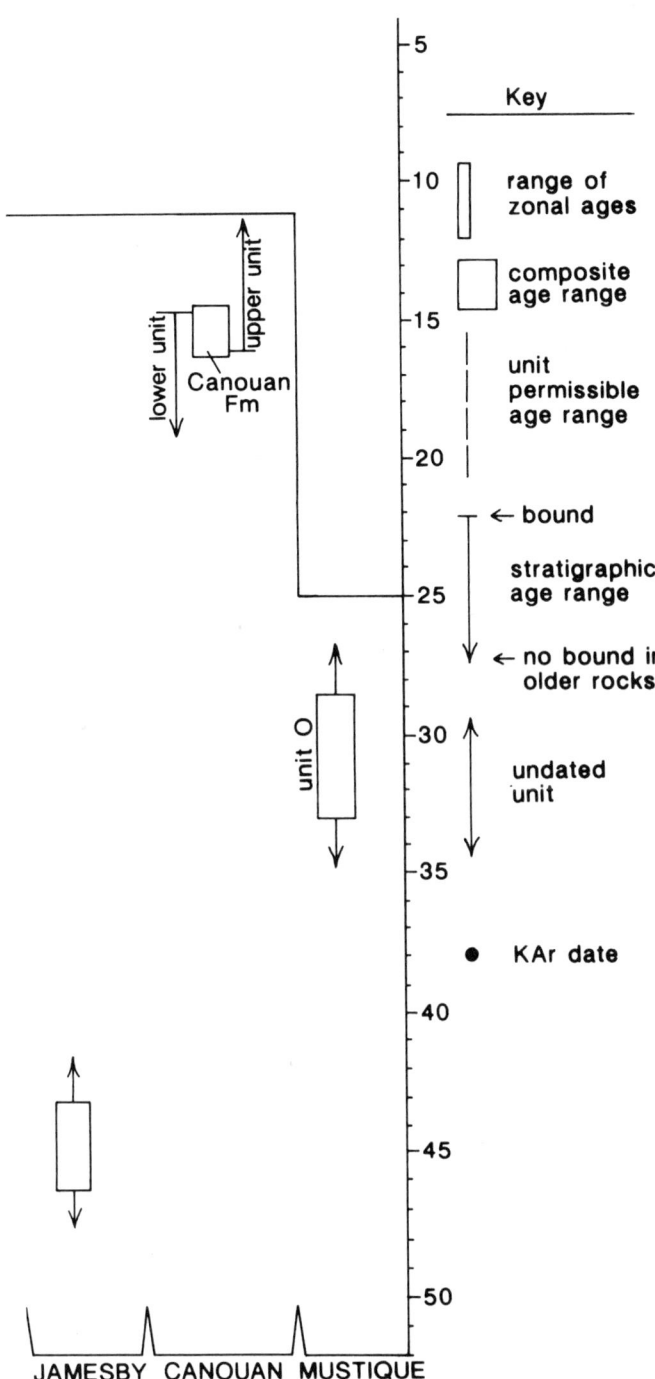

pelagic limestone of Mayreau and Carriacou (Mayreau Basalt and Cherry Hill Basalt, Anse Bandeau Formation, and Bogles Limestone). In both islands, the pillow basalts are the lowest exposed rocks. Our dating shows, however, that the highest pillow basalt and the capping limestone on Carriacou are between 0.5 and 6 m.y. younger, within middle Eocene time, than equivalent horizons in the Mayreau succession (Figs. 27, 28). Nonetheless, we argue that the correlation is correct and that the Mayreau and Cherry Hill Basalts originated in a conterminous unit. We explain their nonsynchroneity in the next section by a sea-floor-spreading origin.

The Baradel chert unit and the chert unit of Mayreau were probably in a contiguous diachronous pelagic deposit with the Anse Bandeau Formation and the Bogles Limestone (Fig. 28). The Baradel chert may have been stratigraphically above the Anse Bandeau Formation, or it may be a complete section of the pelagic deposit from a discrete site.

The Belvedere, Anse La Roche, and Tufton Hall Formations all include upper Eocene beds (Fig. 28) with similar lithic components: arc-volcanigenic clay, sand, and gravel plus marl. The proportions of the lithic components vary markedly among them (Table 5), and the three formations evolved as different facies in a basin-plain-turbidite-fan system. Their paleogeography and initial contiguity are, however, unclear.

The Canouan and Carriacou Formations can be viewed as related by their penecontemporaneity and high contents of clastic carbonate. They were, however, probably not contiguous deposits because the Carriacou is platformal and the Canouan, basinal and turbiditic. It is uncertain whether Carriacou or another mid-Miocene platform was the source of the sediment in the Canouan Formation.

Magmatism

The older rocks of the SLAAP contain Paleogene basaltic magmatic rocks and contain sediments that are products of Paleogene and Neogene magmatism at uncertain sites. These evolved before the onset of Neogene magmatism at each of the sites studied.

The Paleogene magmatic rocks consist of middle Eocene pillow basalt (Mayreau Basalt, Anse Bandeau Formation, Cherry Hill Basalt) and upper Oligocene basalt of the Belvedere Formation of Carriacou. The undated massive porphyry of Union Island is the only siliceous or felsic igneous body of possible Paleogene age recognized in the SLAAP.

The Eocene pillow basalts are the oldest rocks of the SLAAP. Their exposed thickness is estimated at least 500 m on Mayreau, and their subsurface thickness may be much greater. The chemistry of the Mayreau Basalt implies they are nonalkaline and had a spreading origin (Speed and Walker, 1991). By correlation, the Cherry Hill Basalt is a product of the same spreading event. As noted earlier, the middle Eocene spreading event is thought to have been the formation of the oceanic lithosphere of the southern Grenada Basin. The Eocene pillow basalt is not a

clear. The breccia unit of Union Island probably evolved with the sandstone-chert union of Union Island.

Although basalt emerged in the Belvedere Formation of group II in late Oligocene time, its volume is minor relative to that of contemporary strata. This is in contrast to the predominance and pillowed nature of basalt in units of group I; the basalt in the Belvedere is therefore not employed as a correlator.

The most evident lithic correlations of the rock units of the SLAAP are between the pillow basalts and their overlying red

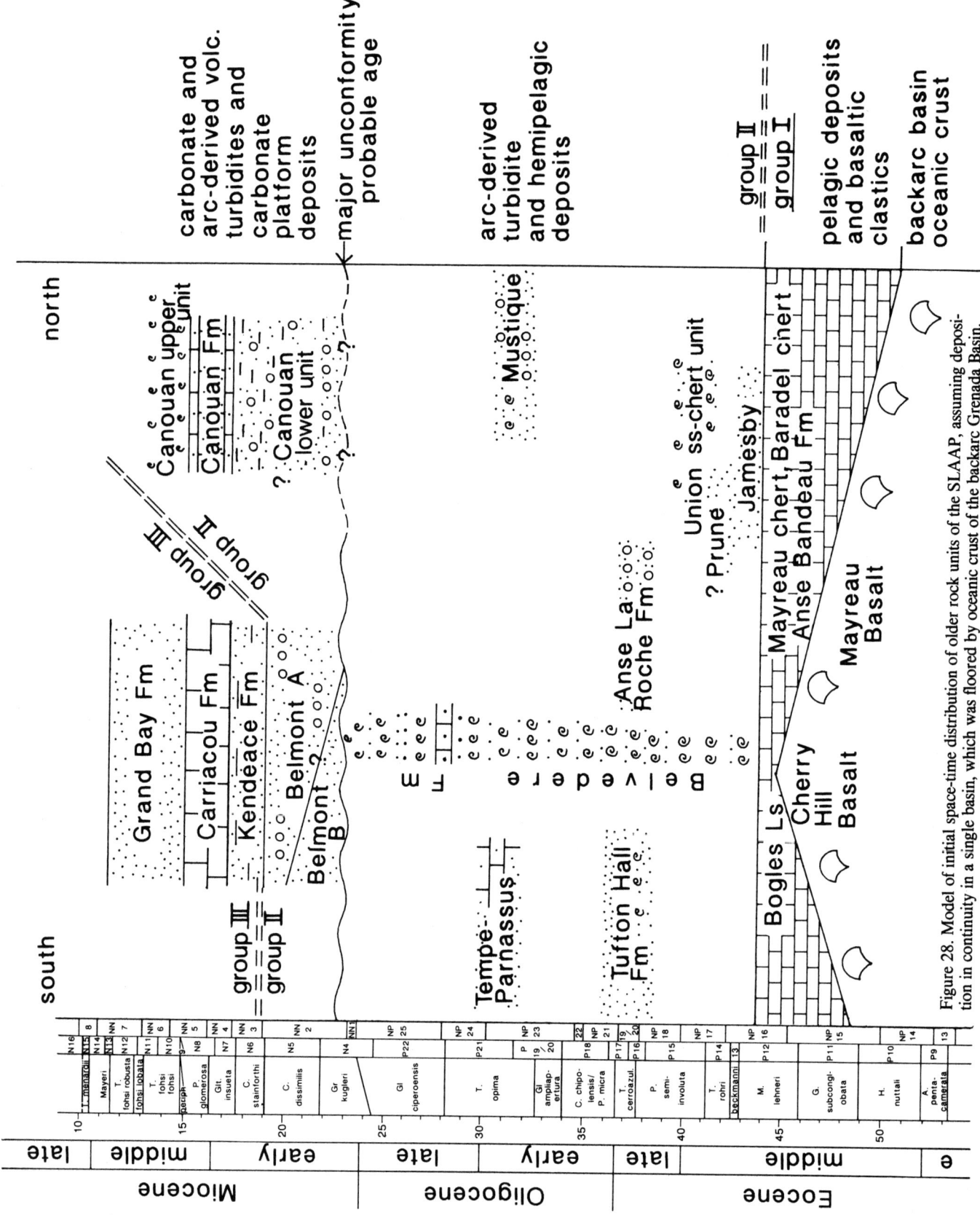

Figure 28. Model of initial space-time distribution of older rock units of the SLAAP, assuming deposition in continuity in a single basin, which was floored by oceanic crust of the backarc Grenada Basin.

product of Paleogene arc magmatism. The age difference between the Mayreau and Cherry Hill Basalts implies that the Cherry Hill lavas emerged at a Grenada Basin spreading ridge between 0.5 and 6 m.y. after the Mayreau Basalt did. Assuming a half spreading rate of 1 cm/yr is a reasonable minimum, a minimum distance of 5 to 60 km is indicated between the initial sites of the two formations. The initial distance, however, is impossible to judge from current positions because the displacements are unknown on major thrusts (Bogles, Baradel), which have transferred the formations, and because the orientation of the spreading center is unknown.

The upper Oligocene basalt of the Belvedere Formation appears to be a product of a short-lived magmatism that vented lapilli and scoria at or just below the seabed at several nearby sites in northeastern Carriacou. Such magmatism was of minor volume relative to penecontemporaneous strata. It may have continued over a longer period, perhaps into early Miocene time if the sub-Belmont lacuna contains eroded Belvedere Formation. The upper Oligocene basalt is of unknown tectonic affiliation, because its extensive sediment contamination and alteration preclude chemical evaluation.

The massive porphyry unit of Union Island is a relatively large, equant felsic intrusion or protrusion that is interpreted to have invaded sediments of the sandstone-chert unit of Eocene age. It is discussed under older rocks because its shape and local spaced foliation distinguish it from the more tabular, albeit similarly altered Neogene magmatic bodies. The foliation is probably related to faulting at the unit contact. This is the only siliceous or felsic igneous body of possible Paleogene age recognized within the SLAAP. Because the massive porphyry of Union Island is undated, however, the possibility is untested. Martin-Kaye (1969) stated that an igneous dike cuts the Tufton Hall Formation near Levera (Fig. 6) and is folded with that formation, implying in-place magmatism before the advent of dated Neogene volcanics (Fig. 5). Our examination of this feature indicates that it is a breccia zone, possibly diatremal, and that it contains no magmatic rock and is not folded. To conclude, we find that the SLAAP includes no nonbasaltic magmatic rocks of demonstrable Paleogene age.

Older sedimentary rock units of the SLAAP contain volcanigenic sediments of two kinds: (1) monolithic basaltic; and (2) arc-derived, with or without basaltic sediment. The distinction of petrographic types is based on the following. Arc-derived is used for rocks with phenocrystic quartz, siliceous-appearing matrix, and content of coarse and (or) abundant phenocrysts of feldspar, clinopyroxene, or hornblende. Basalt is used for dark mafic aphanitic or finely porphyritic rock.

Monolithic basaltic sediment occurs in the Anse Bandeau Formation of group I and upper Belvedere Formation of group II. This is interpreted to have arisen from local vents on the seabed. Sediment-gravity flows of basaltic sediment also occur in part of the Belmont Formation (Belmont B) and in the sedimentary succession of Prune Island, both of group II. The basaltic sediment in these units probably emerged directly from submarine vents somewhere upslope. Such vents were evidently discrete from the sources of arc-derived volcanigenic sediments within these units. The tectonic affiliation of the basalt in the sedimentary units of group II is unknown.

The arc-derived sediments are commonly polymict, and most include rounded gravel, together with biogenic debris of littoral or neritic origin. This implies the source regions were subaerial volcanoes that were fringed by carbonate banks. The freshness of crystal and lithic grains at the time of deposition in volcanigenic strata argues, however, that extrusion in an arc and final transport of debris to the site of deposition were not greatly separated in time. Many of the sediment-gravity flows could have started as pyroclastic flows that swept up earlier deposited particles from volcano flank, beach, and shelf enroute toward a basin floor. Basaltic clasts mixed in with such sediment probably originated from arc lavas.

The oldest recognized arc-volcanigenic sediments of the SLAAP are in the Jamesby rocks and the Belvedere Formation of Carricou. The oldest dated beds in both units are NP16, middle middle Eocene, between about 43 and 46 Ma. In the Jamesby rocks, rounded particles and shallow benthic skeletal debris are absent, indicating the sediment is probably from a nonshoaled source. The earliest well-dated record of sediments from a subaerial, carbonate-fringed arc volcano is late Eocene, in the Anse La Roche Formation of Carriacou and the Tufton Hall Formation of Grenada. If all these units came from the same arc, the change in skeletal content may record the upward growth of the arc from middle to late Eocene time.

The onset in Neogene time of copious local magmatism of the modern arc in the SLAAP was ≤12 Ma, although apparently minor volumes of magma locally emerged back as far as 15 Ma (Fig. 5) and perhaps earlier, if our dates from Mustique validly record Neogene magmatism. Sedimentary units of what we call older rocks were being deposited up until about 13 Ma. A transition thus occurred within the Miocene in volcanigenic sediment provenance from sources outside the SLAAP to sources within it.

To summarize, the first major point of this section is the sequence of volcanic phenomena: middle Eocene extrusion of thick pillow basalt at a spreading center; then accumulation of arc-derived sediment in basinal sites from nonshoaled volcanoes in late middle and possibly, early late Eocene; and finally, deposition of arc-derived sediment in basinal sites from subaerial, carbonate-fringed volcanoes from late Eocene into probably early Miocene time. Second, local venting of basalt occurred in late Oligocene and early Miocene time. Third, there are no rocks recording in-place arc magmatism during Paleogene time in the SLAAP with the possible exception of the undated massive porphyry unit of Union Island.

Sedimentary depositional environments

Sediments of the older rocks of the SLAAP as a whole were deposited in three generally sequential environments, which correspond with stratigraphic groups I to III, discussed above. The environments and their durations are: (I) pelagic basin, middle Eocene; (II) turbidite basin, middle Eocene extending into the middle Miocene; and (III) platform, late early and (or) middle Miocene (Fig. 28). The temporal distinction between turbidite basin and platform is not geographically uniform in the SLAAP. It exists as defined on Carriacou, but, for example, not on Canouan, which continued as a turbidite basin into middle Miocene time. Further, platforms existed outside the SLAAP at unknown sites and were sources of sediment for turbidite basins of the SLAAP as far back as late Eocene.

Pelagic basin. Thin pelagic beds occur in successions in units of stratigraphic group I: Anse Bandeau Formation and chert unit of Mayreau, Baradel chert unit, and Bogles Limestone of Carriacou. These are of known middle Eocene age or are thought to be middle Eocene by lithic correlation. The pelagic beds are either homogeneous or interbedded with pillow lava and basaltic sediment of local origin. These are interpreted to have accumulated as calcareous biogenic ooze. They have subsequently been variably transformed to crystalline micrite and chert by diagenesis and perhaps, by Neogene hydrothermal effects. These pelagic successions represent open-marine deposition above the mid-Eocene carbonate compensation depth (about 3.5 km) in a basin that was distant or barred from clastic debris except for basaltic debris of local origin. The basin floor on which the pelagic succession was deposited is pillow basalt that had evolved by seafloor spreading before and during deposition of the pelagic successions. The abundant red limestone that is interstitial to the basaltic pillows in the Mayreau and Cherry Hill Basalts is interpreted to have been seabed sediment that was intruded and engulfed by the pillowing flows.

Turbidite basin. A turbidite basin was the environment of deposition for the rock units of stratigraphic group II, of middle Eocene to Miocene age. The major deposits in all these units are from sediment-gravity flows of mainly volcanigenic clastics together with varied proportions of biogenic clastic grains. Hemipelagic interbeds are a subordinate to a major constituent. The sediment-gravity flow deposits include turbidites, grain (including gravel) flows, and debris flows. We refer to these collectively as turbidites for economy of terms, even though we differentiate true turbidites as tabular layers with upward fining and related sedimentary structures throughout most of the bed. Some tabular beds with upward fining but no tractional structures may be of ashfall origin.

All of the turbidite basin deposits accumulated below wave-base, based on the following evidence. First, virtually all contain hemipelagic interbeds and angular clasts of pelagic rock that is soft or was so during transport. The beds and clasts contain diverse planktic microfossils but not benthic ones. Second, successions of upward-fining layers are preserved. Depths were evidently subneritic, below about 500 m, above which calcareous benthic forams are thought to have existed, and above the lysocline, at which planktic forams are well preserved.

Among sediment-gravity flows, sandy grain flows consisting of sandstone and (or) pebbly sandstone are most widespread in the older rocks as a whole. True turbidites, as defined above, are somewhat less prevalent, and gravelly grain flows and debris flows are least common.

The grain flows are characteristically thick bedded, massive or plane or steeply cross-laminated, and grain supported. They are stratified by clast orientation and grain-size grading in the upper quarter or less of the flow. Their bases are commonly scoured. These sandy and gravelly deposits were deposited from water-rich, high regime flows in channels. Grain-flow deposits occur together with other types of sediment-gravity flows in the Belvedere, Anse La Roche, and Belmont Formations of Carriacou; units of Prune Island, Jamesby, and Mustique; the Baradel volcanigenic sedimentary rock unit; and the Tufton Hall Formation of Grenada.

Turbidites are most commonly sand rich and base- and top-present or top-absent. These probably represent midfan or outerfan facies in the standard radial fan model. They occur widely in the Belvedere and Tufton Hall Formations and locally, in most of the other older rock units of stratigraphic group II. Sequences of muddy base-absent turbidites, suggestive of basin-plain environments, are scarce except in the Tufton Hall Formation.

Debris flows are unstratified and have a framework of unsorted muddy and sandy sediment and floating coarse clasts. They occur in the nonvolcanigenic unit of Prune Island where we interpret them to have arisen by local resedimentation of muddy and sandy turbidites. They are also present in the unit of older rock of Mustique, the Baradel volcanigenic sedimentary unit, and the Anse La Roche and Belmont Formations of Carriacou.

Chalk, marl, and mudstone occur as discrete thin interbeds in turbidite basin deposits. These probably represent background deposition during intervals between sediment-gravity flows. Their content of planktic flora and fauna with mainly concordant age ranges suggests a pelagic origin rather than one by resedimentation of fines, unless resedimentation occurred rapidly after initial deposition. The sparseness of hemipelagic beds in some units of turbidite basin deposition (e.g., the Anse La Roche Formation of Carriacou) suggests either a high rate of emplacement of sediment-gravity flows or scouring of the seabed by most sediment-gravity flows.

The origin of the marl- and mudclasts, which occur in a fair fraction of sediment-gravity flows regardless of age, is an important question to basin reconstruction. Either they are intraclasts, derived from penecontemporaneous seabed cover within the turbidite basin, or they were picked up in transit between the volcanic source region and the turbidite basin, presumably from a slope and (or) basin floor covered with older pelagic beds. An intraclastic origin seems more likely because (1) the clasts are lithically much like preserved pelagic interbeds; (2) permissible age ranges of the clasts overlap with those of the containing strata, at least for the clasts dated (Fig. 9); and (3) it is easier to conceive than a circumstance where sediment-gravity flows had to transit discrete regions of hemipelagic deposits and (or) concurrent hemipelagic deposition for 25 m.y. or more.

The principal characteristic of the clastic deposits in the turbidite basins is the high regime of the flows that delivered most of them, as indicated by their thickness, grain size, structures, and evidence for internal turbulence. We suspect that most such flows were channelized, deduced from locally scoured bases and content of marl- and mudclasts taken as products of channel-wall collapse, even though exposure lengths are too small to recognize major channels hundreds of meters wide. In contrast, distal turbiditic deposits, either basin plain or interchannel, are sparsely represented. Therefore, we envision deposition on a basin floor in and about major channels that conducted sediment from relatively elevated volcanic platforms that included subaerial regions from late Eocene on. The depositional sites were probably not on the platformal slope because of the absence of abundant muddy rocks and slumps and because the high-regime grain flows would not be expected to stop on a slope.

Owing to the poor and discontinuous exposure of the SLAAP, we are unable to suggest how many basins may be represented among the turbidite-basin deposits. The transport distance is also unknown, perhaps a few kilometers if the deposits lined the base of slope of a backarc escarpment or, on the other hand, many hundreds of kilometers if the delivery system included long aggradational channels like that of the Bengal Fan (Curray and Moore, 1971). The apparently preferred occurrence of debris flows in upper strata of the older rocks (Mustique; Belmont A Formation) may suggest relatively short transport paths in late Oligocene and early Miocene time. Flow direction measurements in the sediment-gravity flows are scattered, due in part to meager data and to deformation. The most consistent set is from steep cross-laminations in the Anse La Roche Formation of Carriacou (Fig. 11e), which indicate southerly flow in today's coordinates.

Platform. The platformal and periplatformal environments for the Kendeace, Carriacou, and Grand Bay Formations of Carriacou are interpreted from the following: (1) concentrations of algal and benthic faunal fragments in massive beds, (2) abundance of whole and articulated megafossils in muddy beds, and (3) copious bioturbation in the Kendeace Formation. The succession of platformal units suggests the emergence in Carriacou of a complex subtidal to supratidal carbonate platform with only minor incursion of volcanigenic sediment in middle Miocene time. The platform was then mainly flooded by volcanigenic sediment in Grand Bay time.

The transition on Carriacou between turbidite basin and platform environments occurred in late early Miocene or early middle Miocene time (Fig. 9), between about 16 and 22 Ma, assuming the Belmont Formation is dated by Belmont B. We are unable to find strata in either the basinal Belmont A or periplatformal Kendeace Formation reflective of the bathymetric transition. Such strata are perhaps a lacuna in the sub-Kendeace unconformity.

The Miocene platformal formations of Carriacou are the only ones recognized among older rocks of the SLAAP. Moreover, the contemporaneous Canouan Formation and perhaps, the lower unit of Canouan were deposited in a turbidite-basin environment, indicating that Miocene vertical motions were nonuniform and that considerable relief existed along strike of what is now the arc platform.

To summarize, the evolution of depositional environments of older rocks of the SLAAP is interpreted as follows. The oldest deposits are pelagic carbonates that were laid down in middle Eocene time in an open oceanic basin during and shortly after extrusion of pillow basalts whose spreading origin probably created the basin. The pelagic basin environment was transformed to a turbidite basin from middle Eocene to middle Miocene time with the influx of arc-derived and possibly other volcanigenic sediment. The turbiditic sediments were delivered mainly by high-regime, probably channelized flows from sources of uncertain positions and distances. The volcanic sources included subaerial regions and carbonate platforms from late Eocene time on. The basinal region transformed to a platformal tract in Carriacou in early or early middle Miocene time, with the deposition of mixed carbonate-volcanigenic sediment facies in shallow-marine and intertidal environments. The region to become the SLAAP was bathymetrically differentiated in mid-Miocene time, however, as turbidite-basin environments persisted at Canouan.

Deformation

In general, the older rocks of the SLAAP are deformed more intensely and pervasively with unit age. Moreover, on some islands, the older rocks contain evident sequential deformations. Our objective in this section is to synthesize a deformation history of the older rocks for the SLAAP as a whole. To do this, we have generalized greatly among scattered exposures, each of which provides only a small sample of kinematic data and control over the time of deformation.

Figure 29 diagrams a deformation history in dimensions of time, geography, and orientation of principal contraction. The contraction orientations, referenced to north, are the horizontal component of the minimum principal elongation, taken as the strike of the XZ strain plane. That plane is inferred either from a single axial plane; from girdles of axial planes, cylindrical bedding, and foliations; or from thrust faults whose attitude and slip direction are known. An arrowhead is included on some bearings to indicate the existence of a consistent direction of vergence or overturning (Fig. 29). In such cases, horizontal simple shear may have been associated with the contraction.

The durations indicated are permissible ranges of each deformation. The maximum age is given by the youngest zonal age of the deformed unit. The minimum age is assigned by the age of a truncative unconformity or by the age of oldest dated local Neogene magmatic rocks. The contraction bearing for each set of structures is placed arbitrarily at the midpoint of the duration; no specific age is implied by this. Each deformation could have persisted over much of the duration, or alternatively, occurred as a single pulse or episodically.

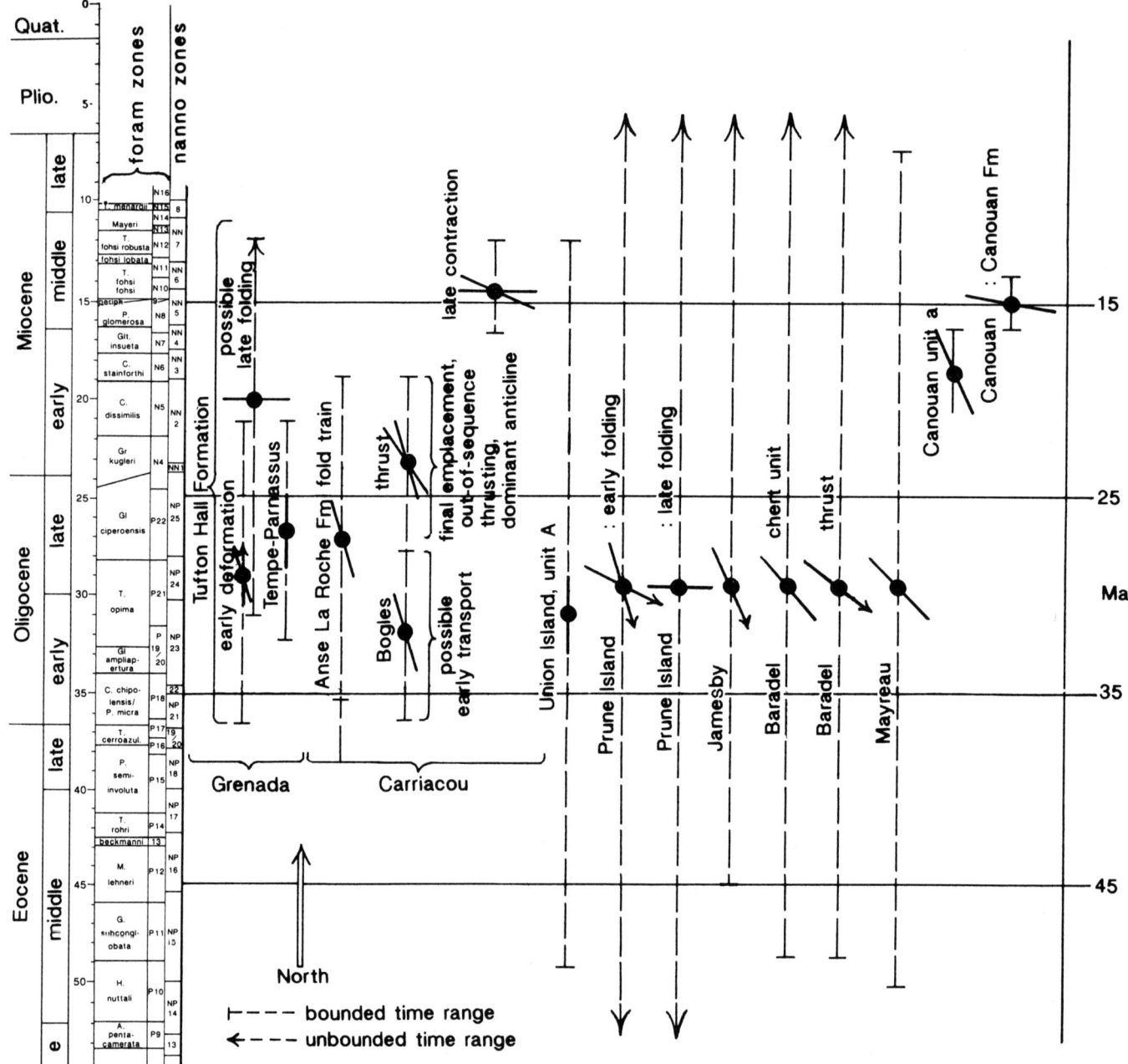

Figure 29. Deformation event history of the SLAAP. Deformation events plotted as the bearing of the horizontal component of the minimum principal elongation as interpreted in text. Arrowheads indicate direction of vergence or overturning of major structures. Dashed lines indicate permissible age ranges of deformation event. Event bearing is arbitrarily placed at midpoint of permissible range. Double bearings indicate range of contraction orientations.

Review of structures. We review here the important structures of the older rocks, emphasizing sequences, interpreted kinematics, and timing. The discussion proceeds from south to north.

On Grenada, the late Eocene Tufton Hall Formation contains early folds and thrusts and disrupted rocks that imply vertical thickening and horizontal shortening on bearings from NS to NNW-SSE (Fig. 6). The vergence of such structures also implies that top-north simple shear on subhorizontal planes accompanied the contractile motions. It is a question, however, whether the north-vergent structures reflect regional kinematics or instead, are parasitic to a more fundamental nonvergent structure at depth, for example, a conjugate thrust system. The middle Oligocene Tempe-Parnassus beds of Grenada are also folded with an orientation indicating north-south contraction. This implies that on

Grenada the north-south contraction occurred in whole or in part after middle Oligocene time, about 20 Ma. We infer that the spatial variability of the early contraction direction is due to rotation by cryptic major open folds due to later east-west contraction. The age of this later event is poorly established, between 30 and about 12 Ma, the latter being the time at which copious Neogene volcanism began on Grenada.

On Carriacou, the principal structures are the train of macroscopic folds of the Anse La Roche Formation, the Bogles thrust that juxtaposes different Eocene facies, the dominant anticline and related thrust within the Bogles allochthon, and folds, faults, and foliation of a later contractile phase. The Anse La Roche fold train implies an origin by buckling during NNW-SSE horizontal shortening (Figs. 10, 11). The individual folds are not fault related, but we infer that the foldtrain reflects deformation of the footwall ahead of the encroaching Bogles thrust. Further, the Anse La Roche foldtrain probably overlies a detachment (unseen) because the folding took place under brittle conditions.

The Bogles thrust brought basement (Eocene basalt) above Eocene sediments (Fig. 10). The correlation of Anse La Roche folding with Bogles thrusting implies either NNW or SSE transport of the Bogles allochthon. The displacement magnitude of the Bogles thrust is unknown but was sufficient to juxtapose mainly basinal and outer-fan strata above contemporaneous inner-fan strata, perhaps a few to a few tens of kilometers. The dominant anticline and related thrust appear to be an out-of-sequence fault-propagation fold that developed in the Bogles allochthon during or after its overriding of Carriacou (Fig. 16). We interpret that the Bogles thrust locally stuck, redeforming the local footwall, and ramped up in a generally northwest direction across the stuck region. The ramp changed upward from a thrust in the basaltic basement rocks to an overturned fold in the suprajacent sedimentary rocks. This out-of-sequence deformation may have been forward- or backward-vergent relative to the transport direction of the Bogles thrust. The out-of-sequence motion occurred between 28 and 16 Ma (late Oligocene to middle Miocene). The Bogles thrust is thus older than 16 Ma and could have had a prolonged movement history. A possible early Oligocene lacuna in the Belvedere Formation might be related to early slip on the Bogles thrust. We speculate that the emplacement of the Bogles allochthon and its out-of-sequence deformation were concluded between about 21 and 23 Ma. This is the probable age of the sub-Belmont unconformity, assuming that Belmont B is a facies of the Belmont Formation and that the Belvedere Formation was seriously eroded during the out-of-sequence movements and development of the unconformity.

The late folds of Carriacou have orientations that indicate contraction with a bearing between EW and WNW-ESE (Figs. 10, 11, 15). One of these folds and a related reverse fault are superposed on the earlier dominant anticline. This phase occurred between late early Miocene (17 Ma) and early late Miocene (11 Ma) times, according to maximum possible zonal ages of strata affected and to known maximum ages of Neogene dikes and normal faults.

Our interpretation of early contraction bearings on Carriacou is based on present mean orientations and the assumption that the change from original attitudes by superposition was small and was achieved by rotation about late axes. If the late deformation includes a component of continuum shortening, the early contraction bearings would have rotated toward the new one where they were not orthogonal.

On Union Island, the deformation of the sandstone-chert unit is interpreted as contractile with north-south bearing (Fig. 17). This is poorly dated within an Oligocene–late Miocene span. The undated older rocks of Prune Island are strongly folded and faulted and have a contraction bearing that varies from NNW-SSE to WNW-ESE across the island (Fig. 18). The vergences are consistently southeast. We explain the change in orientation of structures as due to rotation in a late open synform with north-northeast–trending axial trace (Fig. 18). Therefore, Prune Island, like Carriacou, supplies evidence for sequential contractions.

The fold of Jamesby Island is strongly overturned to the south-southeast (Fig. 19). The existence of poorly developed folded layer-parallel foliation at Jamesby implies sequential deformation but not like that of Carriacou. Similarly, the Baradel chert unit contains layer-parallel foliation and later faulted folds whose orientations indicate northwest-southeast contraction. The Baradel thrust, which underlies the chert unit and may have substantial displacement, had a final direction of transport of S70°E (Fig. 20). Folding of the Baradel chert may reflect hanging wall deformation during the Baradel thrust movement. If so, either the slip direction of the thrust changed during transport or the thrust slip was oblique to hanging wall contraction, which perhaps represents flattening at an oblique ramp. Structures of Baradel are known only to be post–early middle Eocene.

On Mayreau, the principal deformation is the folding of middle Eocene rocks, implying northwest-southeast contraction (Figs. 21, 22). The folding occurred before Neogene magmatism, dated only at 6.8 Ma. The age and direction of faulting of the Mayreau chert unit are unknown. This fault is unlikely to be the Baradel thrust, which placed higher over lower metamorphic or diagenetic grade.

Canouan contains good evidence for sequential contraction before the onset of local Neogene magmatism. The early contraction of the lower unit is approximately NNW-SSE bearing (Fig. 25) and may have been associated with metamorphism and microfoliation development. It occurred before about 16 Ma, the minimum age of the unconformity at the base of the Canouan Formation. Late contraction, which produced the syncline of the Canouan Formation, had a WNW-ESE bearing. This event is younger than 16 Ma and preceded the deposition of the upper unit, before about 7 Ma. We have inferred by sedimentologic linkage that the deposition of the lower unit and the Canouan Formation was related and not greatly separated in time. If so, it follows that the early contraction probably was an early Miocene pulse that rumpled the seabed and caused a submarine unconformity to develop but did not efface the turbidite basin.

Mustique contains shallowly south-dipping homoclinal

strata of middle Oligocene age (Fig. 26). The kinematic origin of the homocline is uncertain, and neither contraction nor extension can be inferred in such rocks. The lack of evident strain on Mustique suggests that the contractions of the other islands demarcate a zone or belt of deformation whose northern margin is between Canouan and Mustique.

Contraction history. Sequential contractile structures are demonstrable on Carriacou and Canouan and are inferred on Grenada and Prune Island. On these islands, the later contraction is westerly (EW to WNW-ESE) whereas the earlier contraction is generally northerly (NS to NW-SE). On islands where only one phase of contraction before Neogene magmatism and extension has been recognized, the bearing of that contraction allies it with the early phase (Fig. 29), except for the Baradel thrust. Therefore, we interpret the data of Figure 29 to indicate that the early northerly contraction affects the older rocks of the SLAAP as far north as at least Canouan. The later westerly contraction may have been local. Alternatively and perhaps, more likely, the later phase was pervasive but with such gentle effects that it cannot be distinguished in small exposure areas of the pre-Miocene rocks, such as those of Baradel, Jamesby, Union, and Mayreau.

The northerly contraction was ongoing in late Oligocene and (or) early Miocene time in Carriacou and Grenada. We suspect the northerly contraction was active in Canouan during the same duration from indirect evidence given earlier. We propose therefore that northerly contraction affected older rocks of the SLAAP as far north as Canouan in late Oligocene and (or) early Miocene time, between about 19 and 28 Ma, assuming Belmont B beds are autochthonous. Except for the possible lacuna within the Belvedere Formation, stratigraphic and structural data provide no evidence for a protracted duration of the northerly contraction.

The later contraction occurred in middle or (and) late Miocene time during or (and) after the heterogeneous uplift of SLAAP from wholly basinal to partly platformal elevations relative to sea level. This is clear because the later contraction affects the platformal Carriacou Formation and the contemporaneous basinal Canouan Formation.

There is no consistent direction of vergence or overturning among older units of the SLAAP. Further, it is questionable whether any of the vergent structures can be interpreted for regional sense of shear, except possibly for the Baradel thrust, which is discussed below. The caution stems from the small exposure volumes of most units and the uncertainty whether unexposed complementary conjugate structures exist.

Deformation Regimes. All the older rocks of the SLAAP have undergone deformation under very low grade metamorphic or submetamorphic conditions. In each unit, deformation included brittle faulting, implying a shallow, low-temperature environment. In some units, however, spaced pressure solution foliations and shear bands exist, commonly in carbonate-rich rocks, and a multilayer fold approaching similar style occurs on Mayreau. Moreover, metamorphic epidote and chlorite exist in the Mayreau Basalt and the lower unit of Canouan. Such features suggest that slight elevation of temperatures occurred locally, perhaps associated with defluidization and phyllosilicate maturation. There are, however, no tectonitic structures and no suggestion of crystal-plastic deformation mechanisms.

In fact, differences in the development of microstructures suggest two deformation regimes among older units of the SLAAP. The first, characterized by moderately developed foliation and lithic evidence of advanced diagenesis or very low grade metamorphism, includes units of the central Grenadines: Eocene rocks of Mayreau, Jamesby, the Baradel chert unit, and the lower unit of Canouan. The second regime is characterized by rare or no foliation and by only slight or local diagenesis. It includes the rocks of Grenada and Carriacou, and the volcanigenic sediment unit of Baradel. The sandstone-chert and breccia units of Union Island and the Prune Island units are ambiguous as to affiliation because they are generally unfoliated but strongly lithified and deformed.

The two regimes probably differ in temperature, hence depth, of deformation. The Baradel thrust is the only exposed contact between rocks of the two regimes. This implies that the higher grade Baradel chert unit was transported from depth above an unmetamorphosed footwall by thrusting, at least partly with east-southeast transport. To generalize, we suggest that the Baradel thrust may be the regional contact between the deeper seated older rocks of the central Grenadines and the shallower older rocks to the east and south. If true, the Baradel thrust may be the surface of greatest displacement in the SLAAP and may have significant throw. Unfortunately, the age of the Baradel thrust is unknown, and we are uncertain whether its transport was during the early or late phase of contraction.

TECTONICS

Six-stage evolution of the SLAAP

We now generalize the preceding information on the older rocks of the SLAAP to a six-stage evolutionary model.

Stage 1: Oceanic crust. Geological development of rocks now exposed in the SLAAP began in early middle Eocene time with the growth of ocean crust by sea-floor spreading. Spreading that created the Mayreau Basalt ended between about 46 and 50 Ma and the Cherry Hill Basalt, between 0.5 and 6 m.y. later. The regional duration of sea-floor spreading before and after these times is uncertain. Pelagic carbonate sedimentation occurred during and immediately following formation of basaltic crust, implying the early middle Eocene seabed was no deeper than about 3.5 km. If the Baradel chert and chert of Mayreau were originally mainly siliceous pelagic deposits that succeeded the pelagic carbonate, the bathymetry of the middle Eocene basin increased, perhaps reflecting thermal contraction of the basaltic lithosphere. The only clastic sediment deposited during stage 1 was basaltic and almost certainly locally derived from edifices of oceanic crust or contemporary eruptions at nearby sites.

Stage 2: Basinal sedimentation and a Paleogene arc. Later in the middle Eocene, between about 43 and 46 Ma, arc-derived volcanigenic sediment together with calcareous hemipelagic sediment began to accumulate on the oceanic crust and its pelagic cover. Such sedimentation and the basinal environment persisted through the Oligocene and at least locally, into middle Miocene time. The seabed was above the carbonate compensation depth for all or most of this duration.

Minor volumes of basalt of uncertain tectonic affinity extruded within the basin in the late Oligocene (Belvedere Formation). There was, however, no magmatism of demonstrable arc origin within the sedimentary basin until the advent of volcanism of the Neogene arc, which was underway at about 14 Ma but occurred mainly after about 12 Ma in Grenada and the Grenadines.

The Paleogene magmatic arc that gave rise to the volcanigenic sediment of the older rocks of the SLAAP has not been located. Its position, orientation, and distance from the basin are unknowns. The existence of shallow-marine skeletal sediment in the volcanigenic turbidites in late Eocene and later time but apparently not in the middle Eocene suggests that the Paleogene arc grew from a submarine to a shoaled edifice in late Eocene time. Further, the Paleogene arc may have begun activity in the middle Eocene.

Stage 3: Northerly contraction. In late Oligocene and (or) early Miocene time, between 16 and 28 Ma, the oceanic crust and its sedimentary cover underwent regional contractile deformation on bearings between north-south and northwest-southeast. The deformation included basement-rooted thrusting and folding by buckling and by fault propagation. Foliations developed in rocks of a probably deeper deformation regime in the central Grenadines but only to a meager degree in an apparently shallower regime in the southern Grenadines and Grenada. Rocks of the two regimes may be juxtaposed regionally by the Baradel thrust. It is uncertain whether that thrust was active in stage 3 or 4. The stage 3 contraction includes no evident uniform sense of horizontal shear.

Stage 4: Uplift and local initial Neogene magmatism. The uplift of the SLAAP relative to sea level as a half-horst is recorded by the development of one or more carbonate platforms above deformed basinal strata in middle Miocene time, about 16 Ma. The beginning of uplift was probably earlier and may have been related to the development of the sub-Belmont unconformity of Carriacou in the early Miocene, between 21 and 23 Ma (Fig. 27), but possibly as far back as late Oligocene time. The height of the uplift relative to sea level is unclear because we have no paleobathymetry for rocks deposited just before uplift began. Moreover, the uplift height was nonuniform in the SLAAP because part of it remained basinal into middle Miocene time or later.

The completion of uplift apparently preceded the arrival in large volume of Neogene arc magma to surface and near-surface levels in the SLAAP by a few million years. It is conceivable, on the other hand, that the beginnings of uplift and advection of Neogene arc magmas were concurrent; such speculation may gain credibility if our early Miocene dates of dikes on Mustique can be shown to be ages of magmatism. If uplift and magma advection began together, the early phase of magmatism, between about 25 and 12 Ma, was almost entirely intrusive and deep-seated.

Stage 5: Westerly contraction. Contraction of stage 5 had bearings between EW and WNW-ENE. It occurred after uplift of the half-horst of stage 4, after about 16 Ma but before 12 Ma, the advent of widespread Neogene volcanism. The westerly contraction had generally small magnitude and apparently developed heterogenously in the SLAAP from Canouan south.

Stage 6: Normal faulting and widespread Neogene volcanism. Normal faulting manifesting apparently minor extensile strain began during the onset of widespread Neogene volcanism in the SLAAP at about 12 Ma. Such volcanism has been volumetrically concentrated on the major islands of the SLAAP: Grenada, St. Vincent, and St. Lucia where it continues today. The magmatism may have extinguished in the Grenadines except for Kick-Em-Jenny by 2 Ma. In general, the attitudes of normal faults and dikes do not indicate a regionally uniform orientation of maximum principal elongation.

Regional correlations and interpretations

We now relate the six-stage history of the SLAAP to features and events of the southeastern Caribbean and suggest regional tectonic interpretations.

Backarc basin. The pre-Miocene older rocks of the SLAAP are thought to have evolved in the backarc Grenada basin, first, by sea-floor spreading in stage 1 and then, by deep-marine sedimentation on the backarc-basin crust in stage 2. The extent and configuration in the Eocene of the backarc oceanic crust are unknown, owing partly to subsequent tectonics. The island arc of the Aves Ridge (Fig. 1) that developed on Venezuelan Basin crust in pre-Eocene time probably became the remnant arc during backarc spreading, and a frontal magmatic arc developed at the southern and (or) eastern side of the basin. The frontal arc, which has not been located, was almost certainly the source of volcanigenic and shallow-marine carbonate clastics that accumulated on the backarc basin crust in stage 2 (middle Eocene to late Oligocene or early Miocene). Thus, we infer that the Paleogene stratal section of the Grenada Basin, which is undrilled, is correlative with that of the SLAAP. If correct, the inference suggests that the transition in the Grenada Basin from tabular reflectors low in the section to mounded and onlapping deposits above (4 to 5s in Fig. 4A; 5 to 6s in Fig. 4B) may correlate with the sub-Belmont unconformity of Carriacou (Fig. 28), which is probably early Miocene, possibly late Oligocene.

Belt of Northerly Contraction. Northerly contraction of stage 3 marked the first deformation of older rocks of the SLAAP, an event of early Miocene and (or) late Oligocene age. The northern margin of this deformation zone is probably between Canouan and Mustique (Fig. 30), but other margins of the zone are not located.

We correlate this zone with a deformation belt south of the Southern Grenada Basin Deformation Front (SGBDF, Figs. 2b, 30) for several reasons. The east-northeast trend of the SGBDF suggests northerly contraction is likely in the belt behind it, and the SGBDF belt and the SLAAP are nearly colinear (Fig. 30). Moreover, the ages of the two may be the same, if the pre-middle Miocene age of the SGBDF, assigned by long-distance seismic stratigraphic correlation by Pinet and others (1985) is valid.

At the southern margin of the Grenada Basin (Fig. 30), the SGBDF is interpreted by Pinet and others (1985, their Fig. 4) as the front of an accretionary prism with northerly imbrication above oceanic crust. From our study of seismic sections across the SGBDF (P. L. Smith and R. Speed, in preparation), we agree and infer that north-south convergence occurred between the Grenada Basin crust and what is now the arc platform west-southwest of Grenada (Fig. 30). The SGBDF prism built up by accretion of strata from the underriding Grenada Basin crust. We have assigned the southern edge of the SGBDF prism (Fig. 30) arbitrarily to the zero free-air gravity locus (Fig. 2c) and have interpreted the prism to continue well southwest of the remnant oceanic crust of the Grenada Basin to a position west of Margarita Island (Fig. 30) on the basis of a narrow belt of negative anomalies where the seabed is close to sea level (Figs. 2c, 30). This southwesterly extension of the prism presumably lies at a locus of total consumption of oceanic crust of the Grenada Basin and above a collisional suture at depth between nonoceanic lithospheres. The suturing in this segment may have arrested convergence throughout the belt. We emphasize that the area of oceanic crust in the Grenada Basin may have been greatly diminished by subduction relative to its late Eocene extent.

From current understanding of the regions underlain by the SGBDF belt and that of the northerly contraction in the SLAAP, an jog of the deformation front is suggested across the western flank of the SLAAP (Fig. 30). This may be due to a local swerve to the north in the strike of the front or to a left-lateral component of slip in the Neogene faulting that raised the SLAAP as a half-horst relative to the Grenada Basin crust.

If the correlation and interpretations above are correct, the northerly contraction of stage 3 of the SLAAP arose by accretion of strata from the relatively south moving, underriding oceanic

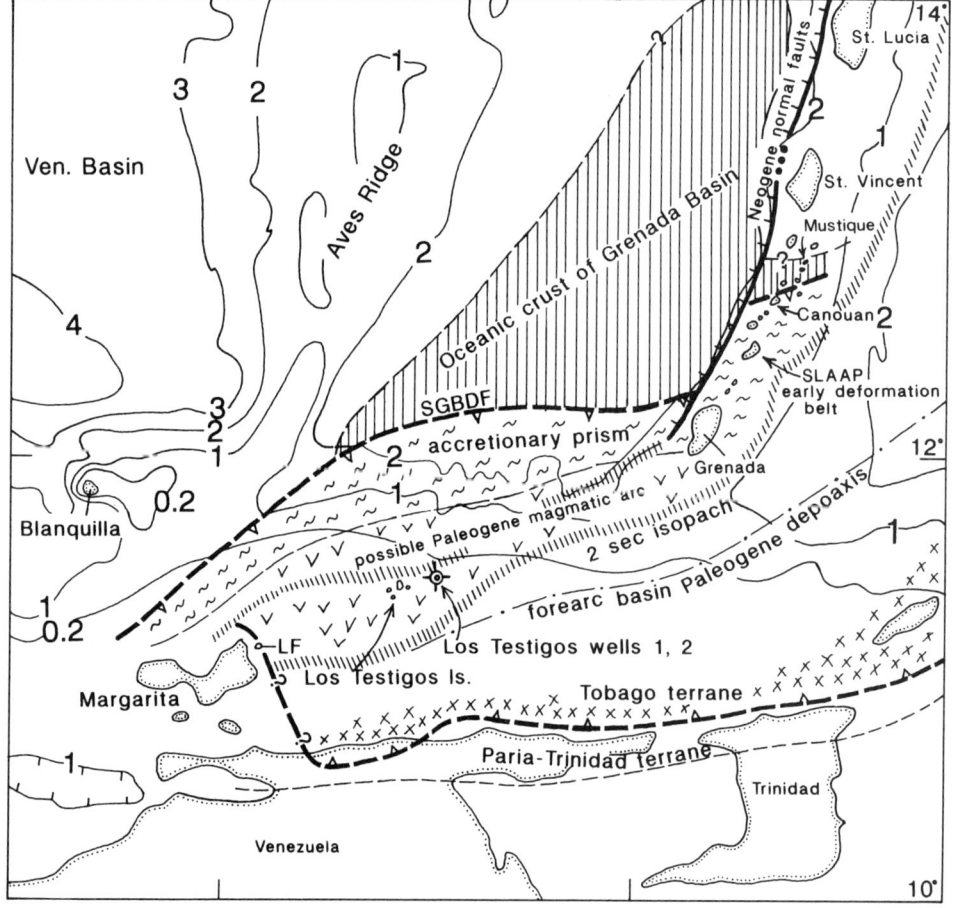

Figure 30. Deduced loci of the Paleogene magmatic arc, early deformation belt of northerly contraction in the SLAAP, and deformation belt (accretionary prism shown by squiggle pattern) bounded by the SGBDF—the southern Grenada Basin deformation front. LF is Los Frailes Islands, x pattern is Tobago terrane, V pattern is possible Paleogene magmatic arc, line pattern is oceanic crust of Grenada Basin. Contours of bathymetry in kilometers.

crust of the Grenada Basin. Further corollaries are as follows: (1) vergence in the early deformation of the SLAAP should have been principally northerly; (2) accretion included one or more slices of oceanic crust in addition to the sedimentary fill of the Grenada Basin; (3) the underthrust oceanic crust of the Grenada Basin probably composes the high-velocity layer that dips south below Grenada (Fig. 3, section X–X′); and (4) the Baradel thrust is a backthrust of stage 3 contraction; or alternatively, a product of stage 5 contraction.

The south-thickening wedge of low-velocity crust (6.2 km/sec) that lies below and south of Grenada (Fig. 3, section X–X′) may include a terrane of crystalline rock, possibly the Tobago terrane, as well as the southern reach of the SLAAP prism. The inference comes from the compressional velocity, probably too fast for even the deep rocks of an accretionary prism, and from the apparent continuity and thickening of the layer to the south. The Tobago terrane may have acted as the backstop for accretion of the SLAAP prism. The crustal boundary between the low- and high-velocity crusts from Grenada south (Fig. 3, section X–X′) is thus inferred to be an inactive thrust fault.

The existence or whereabouts of the SGBDF-SLAAP deformation belt to the east of the SLAAP is uncertain.

Paleogene magmatic arc

Rocks indicative of intrusion or extrusion of Paleogene arc magmas within the Grenadines or Grenada have not been identified, and we infer that the region now occupied by the SLAAP was not a magmatic arc in Paleogene time. The existence of such an arc somewhere in what is now the southeastern Caribbean, however, is clear from the provenance of sediments in Eocene and Oligocene strata of these islands. We now examine where the Paleogene magmatic arc may be.

As noted, the middle Eocene-Oligocene magmatic arc was probably on the south and (or) east side of the Eocene backarc Grenada Basin. This is because the arc on the opposite side, the Aves Ridge, appears to have died early in Eocene time and is onlapped by Grenada Basin strata.

Next, we consider three options for sites of the arc that obey the preceding condition. The first option is that the Paleogene magmatic arc is far to the west of the SLAAP, bypassed by eastward Neogene translation of the Caribbean Plate relative to South America. This possibility requires that Ca-Am relative velocity has been taken up principally in the northern reaches of the Ca-Am plate boundary zone. There, the SGBDF belt is the most evident candidate to have taken up large displacement. Given the earlier rationale that Neogene relative motion has been approximately constant (east-southeast, 1.3 cm/yr), the east-west, right-lateral displacement along the SGBDF might be as much as 300 km in 25 m.y. If the magmatic arc had been bypassed by such a process, its current position would probably be in the ridge west of Blanquilla (Fig. 30) containing the Venezuelan islands, La Orchila and Las Aves. Such islands, however, expose no Paleogene igneous rocks and no evident record of Paleogene heating (Schubert and Moticska, 1973). Therefore, within the limits of poor exposure, we infer the Paleogene arc was not abandoned far to the west of the SLAAP.

Options 2 and 3 assume that Paleogene magmatic arc has moved more or less with the Grenada Basin.

The second option is that the Paleogene magmatic arc is buried within the present arc platform from St. Vincent north (Fig. 30). The thick low-velocity crust of that area is the only feature supporting the hypothesis. No Paleogene magmatic rocks are known in the Neogene arc platform from St. Vincent north for 400 km to Antigua (Le Guen de Kerneizon and others, 1983; Mascle and Westercamp, 1983).

The third option is that the locus of the Paleogene arc platform was along the southwestern leg of the present arc platform, west-southwest of Grenada and east of Margarita (Fig. 30). It is supported by the following evidence. First, rocks exposed and drilled at two sites on the proposed arc are permissive of local Paleogene magmatism. Los Testigos Islands (Fig. 30), the only outcrops between Grenada and the vicinity of Margarita contain volcanic and intrusive rocks that are lightly metamorphosed (Schubert and Moticska, 1973). Three amphibole separates from metagranitic rocks give K-Ar dates of 44 ± 4, 44 ± 5, and 47 ± 6 Ma (Santamaria and Schubert, 1974). Such dates could represent middle Eocene crystallization, although they could alternatively be mixed ages of much older magmatic crystallization and younger partial recrystallization. Igneous dikes cut the granitic rocks, and these may represent a Paleogene event. The Los Testigos wells 1 and 2 (Fig. 30) bottomed in unmetamorphosed volcanigenic rocks said to consist of bombs and lapilli and to be a pyroclastic deposit (Castro and Mederos, 1985). The drilled volcanigenic unit is at least 55 m thick and lies with discordance below lower Miocene beds. Whole-rock K-Ar dates are 39.6 ± 2 and 35.5 ± 2 Ma (Castro and Mederos, 1985). Although admissive, the well unit does not prove local venting because of the difficulty in well cuttings of distinguishing hot fragmental deposits from coarse fresh volcanigenic sediments, for example, that of the contemporaneous Anse La Roche Formation.

Second, evidence for a Paleogene forearc basin exists south of the proposed arc (Fig. 30). This is the Carupano Basin (Fig. 2; Pereira and others, 1985) whose drilled Eocene and Oligocene strata thin both north and south of the depoaxis shown in Figure 30 and include ash and volcanigenic turbidite (Speed and others, 1989).

To conclude, current evidence favors option 3, that the Paleogene magmatic arc lies west-southwest of Grenada, as shown in Figure 30.

The question arises whether or not the Paleogene arc's current orientation (WSW-ENE, Fig. 30) is the same as that of Paleogene time. The paleomagnetic studies that address the question are as follows.

1. *Tobago.* MacDonald and Van Horn (1977) obtained a paleomagnetic direction of 327° for the mean of five sites with good structural control and evidence for field reversals in mid-Cretaceous volcanic rocks. Although site means have large uncertainties, the mean mean direction is precise enough to suggest

counterclockwise rotation of Tobago relative to the South American direction at 100 Ma at Tobago. The paleomagnetic data from diorite on Tobago from Hargraves and Skerlec (1983) have no structural correction and have large scatter.

2. *Mayreau.* Paleomagnetic data from Mayreau Basalt (M. Beck, R. Burmester, and R. Speed, in preparation) discussed earlier (Fig. 22f), indicate either no or small clockwise (<11°) rotation and no latitudinal translation of Mayreau relative to Eocene South America.

3. *Los Testigos and Los Frailes Islands.* Paleomagnetic directions of Hargraves and Skerlec (1983) from oriented hand samples of igneous dikes and plutons from Los Testigos (Fig. 30) have varied northwest and northeast trends, and those from dikes on Los Frailes (LF, Fig. 30) have easterly trends. Both data sets have larger scatter, small sample numbers, and no structural correction.

If the Tobago terrane underlies the Paleogene magmatic arc and has undergone no internal rotation about a vertical axis since the Cretaceous, the paleomagnetic data of MacDonald and Van Horn (1977) might suggest the Paleogene arc was previously more east-west-trending than it is today. Such data, however, should be regarded as inconclusive because the rotation history of the Tobago terrane is unknown and may have entirely preceded the Paleocene arc.

The paleomagnetic data from the Mayreau Basalt suggest little or no translation or rotation relative to Eocene South America. If the Paleogene arc's east-northeast strike arose by rotation during or after underthrusting of Grenada Basin crust and northerly contraction in the SLAAP, the Mayreau Basalt, which represents such crust in the accretionary belt, should probably also be rotated.

To conclude, the Paleogene orientation of the arc is unknown. Paleomagnetic data provide no indication that the arc has undergone large systematic rotation to its current orientation.

We now suggest a plate-tectonic history of the Paleogene magmatic arc and adjacent tracts shown in Figure 30. The Tobago terrane (Figs. 2, 30), whose plate affiliation is unclear, may have attached to the Caribbean Plate before or during the middle Eocene backarc spreading of the oceanic crust of the Grenada Basin. The Tobago terrane became the forearc basement in the Eocene, and its southern margin was the leading edge of crystalline lithosphere of the Caribbean Plate. The forearc thrust above subducting Atlantic lithosphere, which was attached to the South American continent and belonged to the American Plate. The Paleogene velocity vector of the Caribbean Plate relative to the American Plate was perhaps east-southeast. The Paleogene magmatic arc was fueled by the subduction on its southern side, and the Carupano forearc basin developed between the magmatic arc and the trench. The arc system then collided with South America by overrunning the continent's slope and outer shelf probably late in Oligocene time (Speed, 1985). Further consumption of the north-south component of plate motion was taken up by northerly contraction within the arc system during late Oligocene and early Miocene times. Shortening mechanisms were (1) backarc subduction and the development of the SGBDF and SLAAP deformation belts; (2) ramping up of the forearc basement, the Tobago terrane, above the accretionary forearc, the Paria-Trinidad terrane (Fig. 30), whose maximum metamorphism is now known to be at about 25 Ma (Speed and Foland, 1990). This history suggests that magmatism in the Paleogene arc expired during collision, perhaps late in Oligocene time, unless subsequent backarc subduction continued to fuel it. Our interpretation thus views the stage 3 contraction of the SLAAP and the correlative accretion of the SGBDF belt as a short-lived deformation due to collision of the arc system with the South American continent.

Uplift, westerly contraction, Neogene magmatism

In stage 4, the SLAAP underwent uplift relative to sea level as a half horst, beginning in early Miocene time or slightly earlier and finishing in middle Miocene time. The westerly contraction of stage 5 followed the uplift, and extension and the vast bulk of Neogene volcanism occurred after about 12 Ma in stage 6.

Our interpretations of stages 1 and 2 invoke comparable evolutions of the SLAAP in Grenada and the Grenadines and the oceanic realm of the Grenada Basin before late Oligocene or early Miocene time. This implies similar depths to the top of oceanic crust in the two regions before then. Since then, the vertical separation of the two crusts has been close to 12 km. Much of this crustal displacement is clearly due to the tectonic uplift of the SLAAP as a half horst by middle Miocene time. Part of the displacement, however, is due to the difference in sediment loading, large in the Grenada Basin (>5 km, Speed and others, 1984) and small on the SLAAP. This displacement has probably increased steadily through Neogene time.

This early Miocene uplift required volume increase in or (and) below the horst, which we suggest was an inflation by an early (25 to 16 Ma), mainly intrusive phase of Neogene arc magmatism. If magmatism is the correct mechanism for the volume increase, the faulted edge of the half horst lay above a locus of asthenospheric upflow. The half horst cuts across Paleogene structures, implying that such flow was newly generated approximately during early Miocene time. It is further implied that a major kinematic change occurred in the southeastern Caribbean with the demise of the Paleogene arc and the generation of the Neogene arc at a new locus and orientation. We emphasize that the Neogene arc is not a clone of the Paleogene one but reflects a major reconfiguration. Such kinematic change must have been due to a reorganization of plates and plate boundaries in early Miocene time, probably with the development of a more west-dipping Atlantic slab from the SLAAP north. We infer a cause of reconfiguration was the collision of the Paleogene arc system with continental South America.

The westerly contraction of stage 5 affected at least parts of the SLAAP after the half horst had reached its current elevation in middle Miocene time. The tectonics of this episode are unclear. They would seem, however, to be due to stresses applied to the boundary of the horst rather than to the mechanics of magma upwelling, which would intuitively expand the overlying reaches of the horst.

The main phase of surface volcanism of the Neogene arc appears to have begun some 10 to 13 m.y. after the onset of uplift and postulated initial magmatism of the SLAAP. The delay may represent the interval necessary to establish conduits for large discharge through the lithosphere and (or) to transport large volumes of magma and (or) source material in the asthenosphere to the new convergent boundary. Alternatively, the main phase could represent a late change in kinematics of the convergent boundary.

ACKNOWLEDGMENTS

We are grateful to William MacDonald and Hans Avé Lallement for penetrating reviews of this paper, to Judith Gendlin Jones for participation in early phases of our research in the SLAAP, to Trevor Jackson for supplying his unpublished geologic map of Carriacou, and to K. A. Foland for $^{40}Ar/^{39}Ar$ analyses. This study was supported by NSF Grant EAR 8803633.

REFERENCES CITED

Andreieff, P., 1985, Stratigraphic range of Caribbean larger foraminifera from Oligocene to Pliocene: State of knowledge in 1985: Editions Technip, p. 99–100.

Arculus, R. J., 1976, Geology and geochemistry of the alkali basalt-andesite of Grenada, Lesser Antilles island arc: Geological Society of America Bulletin, v. 87, p. 612–624.

Arculus, R. J., and Wills, K.J.A., 1980, The petrology of plutonic blocks and inclusions from the Lesser Antilles island arc: Journal of Petrology, v. 21, p. 743–799.

Argus, D. F., 1990, Current plate motions and crustal deformation [Ph.D. thesis]: Evanston, Illinois, Northwestern University, 163 p.

Avé Lallement, H. G., and Guth, L. R., 1991, Role of extensional tectonics in exhumation of ecologites and blueschists in an oblique subduction setting, northeastern Venezuela: Geology, v. 18, p. 950–953.

Bellizzia, A., 1985, Sistema montanosa del Caribe—una Cordillera aloctona en la parte norte de America del Sur: Proceedings VI Congresso Geologica Venezolano, v. 10, p. 6657–6835.

Berggren, W. A., Kent, D. V., Flynn, J. J., and Van Couvering, J. A., 1985, Cenozoic geochronology: Geological Society of America Bulletin, v. 96, p. 1407–1418.

Biju-Duval, B., Mascle, A., Montadert, L., and Wanneson, J., 1978, Seismic investigations in the Columbia, Venezuela, and Grenada Basins and on the Barbados Ridge for Future IPOD drilling: Geologie en Mijnbouw, v. 57, p. 105–116.

Bladier, I., 1977, Rocas verdes de la region de Carupano, Venezuela: Caracas Boletin Geodinimica, Comité International de Geodinimica, Grupo 2, p. 35–49.

Blow, W. H., 1969, Late middle Eocene to Recent planktonic foraminiferal biostratigraphy, in Bronnimann, R., and Renz, H. H., eds., Proceedings of 1st International Conference on Planktonic Microfossils, Geneva, 1967: Leiden, E. J. Brill, p. 199–421.

Bolli, H. M., Saunders, J. B., and Perch-Nielsen, K.v.S., eds., 1985, Plankton stratigraphy: Cambridge, Cambridge University Press, 401 p.

Bouysse, Ph., 1988, Opening of the Grenada back-arc basin and evolution of the Caribbean Plate during the Mesozoic and early Paleogene: Tectonophysics, v. 149, p. 121–143.

Bouysse, P., Andreieff, P., Richard, M., Baubron, J. C., Mascle, A., Maury, R. C., and Westercamp, D., 1985, Aves Swell and northern Lesser Antilles Ridge: Rock-dredging results from Arcante 3 cruise, in Mascle, A., ed., Symposium Geodynamique des Caraibes: Paris, Technip, p. 65–76.

Bowin, C., 1976, Caribbean gravity field and plate tectonics: Geological Society of America Special Paper, v. 169, 76 p.

Boynton, C. H., Westbrook, G. K., Bott, M.H.P., and Long, R. E., 1979, A seismic refraction investigation of crustal structure beneath the Lesser Antilles island arc: Geophysical Journal of the Royal Astronomical Society, v. 58, p. 371–393.

Briden, J., Rex, D. C., Faller, A. M., and Tomblin, J. F., 1979, K-Ar geochronology and paleomagnetism of volcanic rocks in the Lesser Antilles island arc: Philosophical Transactions Royal Society of London, Series A, v. 291, p. 485–528.

Bunce, E. T., Phillips, J. D., Chase, R. L., and Bowin, C. O., 1970, The Lesser Antilles arc and eastern margin of the Caribbean Sea, in Maxwell, A. E., ed., The Sea, volume 4: New York, Wiley Interscience, p. 359–385.

Butterlin, J., 1981, Claves para la determinacion de macroforaminifera de Mexico y del Caribe, del Cretacio superior al Mioceno medio: Instituto Mexicano del Petroleo, 218 p.

Castro, M., and Mederos, A., 1985, Litoestratigraphia de la Cuenca de Carupano: Proceedings VI Congresso Geologica Venezolano, p. 202–225.

Clark, T. F., Korgen, B. J., and Best, D. M., 1978, Heat flow in the eastern Caribbean: Journal of Geophysical Research, v. 83, p. 5883–5891.

Cole, W. S., 1958, Larger foraminifera from Carriacou, British West Indies: Bulletin of American Paleontology, v. 38, p. 327–337.

Curray, J. R., and Moore, D. G., 1971, Growth of the Bengal deep-sea fan denudation of the Himalayas: Geological Society of America Bulletin, v. 82, p. 563–572.

DeMets, C., Gordon, R. G., Argus, D. F., and Stein, S., 1990, Current plate motion: Geophysical Journal International, v. 95, p. 21931–23948.

Diehl, J. F., Beck, M., Beske-Diehl, S., Jackson, D., and Hearn, B. C., 1983, Paleomagnetism of the Late Cretaceous–early Tertiary north-central Montana alkalic province: Journal of Geophysical Research, v. 88, p. 10593–10611.

Edgar, N. T., Ewing, J. I., and Hennion, J., 1971, Seismic refraction and reflection in the Caribbean Sea: American Association of Petroleum Geologists Bulletin, v. 55, p. 838–870.

Engebretson, D. C., Cox, A., and Gordon, R. G., 1985, Relative motions between oceanic and continental plates in the Pacific Basin: Geological Society of America Special Paper, v. 206, 59 p.

Fox, P. J., and Heezen, B. C., 1975, Geology of the Caribbean crust, in The Ocean Basins and Margins, volume III, Nairn, A.E.M., and Stehli, F. G., eds., New York, Plenum, p. 421–466.

Frost, C. D., and Snoke, A. W., 1989, Tobago, West Indies, a fragment of a Mesozoic oceanic island arc: Petrochemical evidence: Journal of the Geological Society of London, v. 146, p. 953–965.

Hargraves, R., and Skerlec, G. M., 1983, Paleomagnetism of some Cretaceous-Tertiary igneous rocks on Venezuelan offshore islands, Netherland Antilles, and Tobago: Transactions, Caribbean Geological Conference, 9th, Santa Domingo, p. 231–238.

Jackson, T., 1970, Geology and petrology of the volcanic rocks of Carriacou, Grenadines [Ph.D. thesis]: Kingston, Jamaica, University of West Indies, 102 p.

Jordan, T. H., 1975, The present-day motions of the Caribbean Plate: Journal of Geophysical Research, v. 80, p. 4433–4439.

Kearey, P., 1974, Gravity and seismic reflection investigations into the crustal structure of the Aves Ridge, eastern Caribbean: Geophysical Journal of the Royal Astronomical Society, v. 38, p. 435–448.

Kearey, P., Peter, G., and Westbrook, G. K., 1975, Geophysical maps of the eastern Caribbean: Journal of the Geological Society of London, v. 131, p. 311–321.

Kennett, J. P., 1978, The development of planktonic biogeography in the Southern Ocean during the Cenozoic: Marine Micropaleontology, v. 3, p. 301–345.

Klitgord, K. D., and Schouten, H., 1986, Plate kinematics of the central Atlantic, in Vogt, P. R., and Tucholke, B. E., eds., The Western North Atlantic Region, M: The Geological Society of America, p. 351–378.

Ladd, J. W., 1976, Relative motion of South America with respect to North America and Caribbean tectonics: Geological Society American Bulletin, v. 87, p. 969–976.

Le Guen de Kerneizon, M., Bellon, H., Carron, J. P., and Maury, R., 1983, L'ile de Sante Lucie: distinction des principales series magmatique á partir des donnes geochronologique: Bulletin Société géologie France, v. 25, p. 845–853.

MacDonald, W. D., and Van Horn, J., 1977, Paleomagnetism of the Hawk's Bill formation, Tobago: Proceeding V Congress Geologica Venezolano, p. 817–834.

Malfait, B. T., and Dinkelman, M. G., 1972, Circum-Caribbean tectonic and igneous activity and the evolution of the Caribbean Plate: Geological Society American Bulletin, v. 83, p. 251–272.

Martin-Kaye, P.H.A., 1958, The geology of Carriacou: Bulletin of American Paleontologists, v. 38, p. 311–324.

—— , 1969, A summary of the geology of the Lesser Antilles: Overseas Geology and Mineral Resources, v. 10, p. 172–206.

Martini, E., 1971, Standard Tertiary and Quaternary calcareous nannoplankton zonation, in Farinacci, A., ed., Proceedings of 2nd Planktonic Conference, Roma: Ediziona Technoscienza, p. 739–785.

Mascle, A., and Westercamp, D., 1983, Geologic d'Antigua, Petites Antilles: Bulletin of Societe Geologie of France, v. 25, p. 855–866.

Maxwell, J. C., 1948, Geology of Tobago, British West Indies: Geological Society of America Bulletin, v. 59, p. 801–854.

Molnar, P., and Sykes, L. R., 1969, Tectonics of the Caribbean and Middle America regions from focal mechanisms and seismicity: Geological Society of America Bulletin, v. 80, p. 1639–1648.

Officer, C. B., Ewing, J. I., Hennion, J. F., Harkrider, D. G., and Miller, D. E., 1959, Geophysical investigations in the eastern Caribbean: Venezuela Basin, Antilles island arc, and Puerto Rico Trench, Physics and Chemistry of the Earth, v. 3, p. 17–109.

Palmer, A. R., 1983, The decade of North American geology, 1983 time scale: Geology, v. 11, p. 504–509.

Pereira, J. G., Perdomo, J. L., and Nelson, M., 1985, Interpretacion sismo-estratigrafica del sector oriental de la cuenca de Carupano: Proceedings VI Congress Geologica Venezolano, Caracas, p. 454–464.

Perez, O. J., and Aggarwal, Y. P., 1981, Present-day tectonics of the southeastern Caribbean and northeastern Venezuela: Journal of Geophysical Research, v. 86, p. 10791–10804.

Pindell, J. L., Cande, S. C., Pitman, W. C., III, Rowley, D. B., Dewey, J. F., LaBrecque, J., and Haxby, W., 1988, A plate-kinematics framework for models of Caribbean evolution: Tectonophysics, v. 155, p. 121–138.

Pinet, B., Lajat, D., LeQuellec, P., and Bouysse, P., 1985, Structure of Aves Ridge and Grenada Basin from multi-channel seismic data, in Mascle, A., ed., Symposium Geodymanique des Caraibes: Paris, Technip, p. 53–64.

Robertson, P., and Burke, K., 1989, Evolution of the southern Caribbean Plate boundary, vicinity of Trinidad and Tobago: American Association of Petrological Geologists Bulletin, v. 73, p. 490–509.

Robinson, E. A., and Jung, P., 1972, Stratigraphy and age of marine rocks, Carriacou, West Indies: American Association of Petroleum Geologists Bulletin, v. 56, p. 114–217.

Rosencrantz, E., Ross, M. I., and Slater, J. G., 1988, Age and spreading history of the Cayman Trough as determined from depth, heat flow, and magnetic anomalies: Journal of Geophysical Research, v. 93, p. 2141–2157.

Rowley, K. C., 1978, Volcanic history of St. Vincent [Ph.D. thesis]: Kingston, Jamaica, University of West Indies, 123 p.

Rowley, K. C., and Roobol, J. M., 1978, Geochemistry and age of Tobago igneous rocks: Geologie en Mijnbouw, v. 57, p. 315–318.

Russo, R. M., and Speed, R. C., 1992, Oblique collision and tectonic wedging of the South American continent and Caribbean terranes: Geology, v. 20, p. 447–450.

Russo, R., Speed, R., and Okal, E. A., 1993, Seismicity and tectonics of the southeastern Caribbean: Journal of Geophysical Research (in press).

Santamaria, F., and Schubert, C., 1974, Geochemistry and geochronology of the southern Caribbean–northern Venezuela Plate boundary: Geological Society of America Bulletin, v. 85, p. 1085–1098.

Saunders, J. B., Bernoulli, D., and Martin-Kaye, P.H.A., 1985a, Late Eocene deep water clastics in Grenada, W. I., ed., Transactions, Latin American Geological Congress, 4th: Trinidad and Tobago, Trinidad and Tobago Printing and Packaging, Ltd., p. 909–918.

Saunders, J. B., Bernoulli, D., and Martin-Kaye, P.H.A., 1985b, Late Eocene deep-water clastics in Grenada, West Indies: Eclogae geologische Helvetica, v. 78, p. 469–485.

Schubert, C., and Moticska, P., 1973, Reconocimiento Geologico de las islas Venezolanas en el mar Caribe entre Los Roques y Los Testigos: Acta Cientificos Venezolanas, v. 24, p. 19–31.

Shepherd, J. B., Rowley, K. C., Beckles, D. M., and Lynch, L. L., 1990, Contemporary seismicity of the Trinidad and Tobago region: Tectonic and earthquake hazard implication, in Proceedings of Conference on Caribbean Seismic Hazards: Mona, Jamaica, University of West Indies (in press).

Snoke, A. W., Yule, J. D., Rowe, D. W., Wadge, G., and Sharp, W. D., 1990, Stratigraphic and structural relationships on Tobago and some tectonic implications: Transactions of the 12th Caribbean Geology Conference, St. Croix, p. 389–403.

Speed, R. C., 1985, Cenozoic collision of the Lesser Antilles arc and continental South America and the origin of the El Pilar fault: Tectonics, v. 4, p. 41–69.

Speed, R., and Foland, K. A., 1990, Mid-Tertiary ^{40}Ar/^{39}Ar ages of metamorphism of Northern Range schists, Trinidad: Second Geological Conference of the Geological Society of Trinidad and Tobago, Port of Spain, Trinidad, April 3–8, 1990, Abstract volume, p. 19–20.

Speed, R. C., and Larue, D. K., 1985, Tectonic evolution of Eocene turbidites of Grenada, in Mascle, A., eds., Symposium Geodymanique des Caraibes: Paris, Editions Technip, p. 101–108.

Speed, R., and Walker, J. A., 1991, Oceanic crust of the Grenada Basin in the southern Lesser Antilles arc platform: Journal of Geophysical Research, v. 96, p. 3835–3852.

Speed, R. C., and 18 others, 1984, Lesser Antilles arc and adjacent terranes, in Atlas 10, Ocean Margin Drilling Program, Regional Atlas Series 28: Woods Hole, Massachusetts, Marine Science International, 28 p.

Speed, R., Torrini, R., Jr., and Smith, P. L., 1989, Tectonic evolution of the Tobago Trough forearc basin: Journal of Geophysical Research, v. 94, p. 2913–2936.

Speed, R., Russo, R., Weber, J., and Rowley, K. C., 1991, Evolution of southern Caribbean Plate boundary vicinity of Trinidad and Tobago: Discussion: American Association of Petroleum Geologists Bulletin, v. 75, p. 1789–1794.

Stein, S., Engeln, J. F., Wiens, D. A., Fujita, K., and Speed, R. C., 1982, Subduction seismicity and tectonics in the Lesser Antilles arc: Journal of Geophysical Research, v. 87, p. 8642–8664.

Sykes, L. R., Kafka, W. R., and McCann, W., 1982, Motion of Caribbean Plate during last 7 million years and implication for earlier Cenozoic movements: Journal of Geophysical Research, v. 87, p. 10656–10676.

Tomblin, J. F., 1975, The Lesser Antilles and Aves Ridge, in Nairn, A., and Stehli, F. G., eds., The Ocean Basin and Margins, 3: New York, Plenum, p. 467–500.

Torrini, R., Jr., and Speed, R., 1989, Tectonic wedging in the forearc basin-accretionary prism transition, Lesser Antilles forearc: Journal of Geophysical Research, v. 94, p. 10813–10859.

Wadge, G., and Shepherd, J. B., 1984, Segmentation of the Lesser Antilles subduction zone: Earth and Planetary Science Letters, v. 71, p. 297–304.

Walker, R. G., 1984, Facies models: Geoscience Canada Reprint Series, v. 1, 317 p.

Westbrook, G. K., 1975, The structure of the crust and upper mantle in the region of Barbados and the Lesser Antilles: Geophysical Journal of the Royal Astronomical Society, v. 43, p. 201–242.

Westercamp, D., Andreieff, P., Bouysse, P., Mascle, A., and Baubron, J. C., 1985, The Grenadines, Southern Lesser Antilles, I, Stratigraphy and volcano-structural evolution, in Mascle, A., eds., Symposium Géodynamique des Caraibes: Paris, Technip, p. 109–118.

MANUSCRIPT ACCEPTED BY THE SOCIETY AUGUST 3, 1992

Printed in U.S.A.

Typeset by WESType Publishing Services, Inc., Boulder, Colorado
Printed in U.S.A. by Malloy Lithographing, Inc., Ann Arbor, Michigan